AF813505

Advances in Experimental Medicine and Biology

Volume 1012

Editorial Board
IRUN R. COHEN, *The Weizmann Institute of Science, Rehovot, Israel*
ABEL LAJTHA, *N.S. Kline Institute for Psychiatric Research, Orangeburg, NY, USA*
JOHN D. LAMBRIS, *University of Pennsylvania, Philadelphia, PA, USA*
RODOLFO PAOLETTI, *University of Milan, Milan, Italy*
NIMA REZAEI, *Children's Medical Center, Tehran University of Medical Sciences, Tehran, Iran*

More information about this series at http://www.springer.com/series/5584

Takeo Kubota • Hideoki Fukuoka
Editors

Developmental Origins of Health and Disease (DOHaD)

From Biological Basis to Clinical Significance

Springer

Editors
Takeo Kubota
Faculty of Child Studies
Seitoku University
Matsudo, Chiba, Japan

Hideoki Fukuoka
Research Organization for Nano & Life Innovation
Waseda University
Tokyo, Japan

ISSN 0065-2598 ISSN 2214-8019 (electronic)
Advances in Experimental Medicine and Biology
ISBN 978-981-10-5525-6 ISBN 978-981-10-5526-3 (eBook)
https://doi.org/10.1007/978-981-10-5526-3

Library of Congress Control Number: 2018944390

© Springer Nature Singapore Pte Ltd. 2018
This work is subject to copyright. All rights are reserved by the Publisher, whether the whole or part of the material is concerned, specifically the rights of translation, reprinting, reuse of illustrations, recitation, broadcasting, reproduction on microfilms or in any other physical way, and transmission or information storage and retrieval, electronic adaptation, computer software, or by similar or dissimilar methodology now known or hereafter developed.
The use of general descriptive names, registered names, trademarks, service marks, etc. in this publication does not imply, even in the absence of a specific statement, that such names are exempt from the relevant protective laws and regulations and therefore free for general use.
The publisher, the authors and the editors are safe to assume that the advice and information in this book are believed to be true and accurate at the date of publication. Neither the publisher nor the authors or the editors give a warranty, express or implied, with respect to the material contained herein or for any errors or omissions that may have been made. The publisher remains neutral with regard to jurisdictional claims in published maps and institutional affiliations.

Printed on acid-free paper

This Springer imprint is published by the registered company Springer Nature Singapore Pte Ltd.
The registered company address is: 152 Beach Road, #21-01/04 Gateway East, Singapore 189721, Singapore

Preface

We have the great pleasure of presenting the first book entitled *Developmental Origins of Health and Disease (DOHaD): From Biological Basis to Clinical Significance*. The editors sincerely appreciate the contributions from all of the authors and their valuable reviews on the theme of this book.

Recent epidemiological studies conducted in various countries, such as England, the Netherlands, and China, suggested that early life deprivation leads to cardiovascular, metabolic, and mental disorders later in life. The DOHaD concept now encompasses a wider scope of early life exposures and a longer list of health outcomes at different life stages. Despite substantial growth in this field of study, major challenges remain, including: (1) identifying the environmental factors responsible for the origins of noncommunicable disorders in fetuses and infants, (2) clarifying the biological mechanisms involved in how environmental factors ultimately cause adult disorders in fetuses and infants, and (3) developing biological markers that can be used during infancy to predict adult disorders in infants exposed to such environmental factors.

Recent studies have also reported that the number of patients with environmental factor-induced noncommunicable disorders, such as diabetes mellitus, is increasing not only in Western countries but also in Eastern countries, such as Asia, the South East, and Africa. Environmental effects during early development can accelerate this trend, resulting in decreased economic power. Therefore, advances in DOHaD studies are important not only for natural science but also for social science.

We hope that this collection of articles will help readers gain a better understanding of the biological mechanisms of DOHaD that affect cardiovascular, metabolic, and neurological systems. We also hope that some of the readers will contribute to the field of DOHaD and make the world healthier.

Chiba, Japan — Takeo Kubota
Tokyo, Japan — Hideoki Fukuoka

Contents

Part I Biological Basis

One-Carbon Metabolism and Lipid Metabolism in DOHaD **3**
Hideoki Fukuoka and Takeo Kubota

Novel Models of Epigenetic Gene Regulation in the Nutritional Environment **11**
Kazuki Mochizuki, Natsuyo Hariya, and Takeo Kubota

Epigenetic Switching and Neonatal Nutritional Environment **19**
Koshi Hashimoto and Yoshihiro Ogawa

Part II Disease and Environment

Developmental Origins of Nonalcoholic Fatty Liver Disease (NAFLD) **29**
Hiroaki Itoh and Naohiro Kanayama

Fetal Origins of Hypertension **41**
Yuichiro Arima, Koichi Nishiyama, Yasuhiro Izumiya, Koichi Kaikita, Seiji Hokimoto, and Kenichi Tsujita

Involvement of Noncoding RNAs in Stress-Related Neuropsychiatric Diseases Caused by DOHaD Theory **49**
Takahiro Nemoto and Yoshihiko Kakinuma

Part III Transgenerational Mechanism and Its Consequences

Placental Development and Nutritional Environment **63**
Kosuke Taniguchi, Tomoko Kawai, and Kenichiro Hata

Risk of Neurodevelopmental Disease by Paternal Aging: A Possible Influence of Epigenetic Alteration in Sperm **75**
Ryuichi Kimura, Kaichi Yoshizaki, and Noriko Osumi

Part IV Clinical Significance

Preemptive Epigenetic Medicine Based on Fetal Programming **85**
Takeo Kubota

Part I

Biological Basis

One-Carbon Metabolism and Lipid Metabolism in DOHaD

Hideoki Fukuoka and Takeo Kubota

Abstract

The predisposing factors to lifestyle-associated diseases are established in the early period of life with underlying gene-environment interaction. Epigenetics is a chemical modification-based genetic mechanism that is affected by various nutritional factors. One-carbon metabolism is a metabolic system associated with methyl residue that is supplied from folic acid. Therefore, from the epigenetic point of view, proper intake of folic acid is important for pregnant women not only to prevent congenital abnormalities such as neural tube defect but also to prevent various adult disorders of the offspring. Dyslipidemia is an important risk factor of coronary heart disease, and epidemiological studies on Dutch winter famine, Jewish holocaust survivors, and Chinese famine suggested that prenatal malnutrition was associated with the dyslipidemia. Recent animal studies revealed that malnutrition in utero causes an epigenetic change in the *Pparα* gene, which accelerates the activity of delta-6 desaturase and delta-5 desaturase, that potentially induces dyslipidemia in adulthood. It has been known that overnutrition also increased the risk of cardiovascular diseases. Recent animal studies revealed that high-fat diet increased DNA methylation in the promoter region of delta-6 desaturase gene (*Fads 2*) that downregulates the gene expression in the arterial smooth muscle, which potentially contributes to cardiovascular diseases. Taken together, either insufficient or excessive nutrition alters epigenetic modification of genes that encodes enzymes associated with lipid metabolism. This altered epigenetic state persists during one's lifetime, which is potentially involved in noncommunicable diseases in adulthood.

Keywords

One-carbon metabolism · Lipid metabolism · DOHaD · Epigenetics · Fads 2

1 Introduction

A number of epidemiological studies provided evidence that support the concept of developmental origins of health and disease (DOHaD). Recent animal studies revealed that malnutrition affects epigenetic modifications, and this is a potential molecular mechanism for DOHaD. These diseases are caused via the two stages, first and sec-

H. Fukuoka (✉)
Research Organization for Nano & Life Innovation, Waseda University, Tokyo, Japan
e-mail: hideoki.fukuoka@gmail.com

T. Kubota
Faculty of Child Studies, Seitoku University, Matsudo, Chiba, Japan

© Springer Nature Singapore Pte Ltd. 2018

T. Kubota, H. Fukuoka (eds.), *Developmental Origins of Health and Disease (DOHaD)*, Advances in Experimental Medicine and Biology 1012, https://doi.org/10.1007/978-981-10-5526-3_1

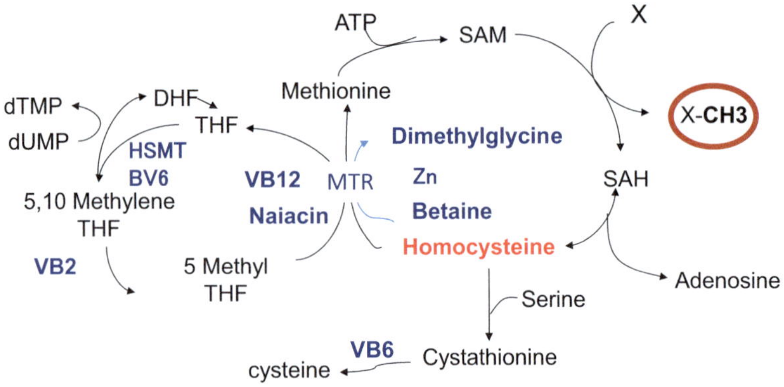

Fig. 1 One-carbon metabolism

ond insults. The predisposing factors to lifestyle-associated diseases are established in the early period of life, such as periconceptional, embryonic, fetal, and neonatal periods with underlying gene-environment interaction (first insult). If individuals with epigenetic changes during fetal period induced by malnutrition have an unideal lifestyle, such as taking excessive nutrition or sedentary lifestyle after birth (second insult), they may develop lifestyle-associated diseases [1].

Epigenetic modifications include DNA methylation, histone modifications, chromatin rearrangements, and noncoding RNAs. Methyl residue is an essential substrate to chemical modification of DNA and histone proteins. One-carbon metabolism is a metabolic system associated with the methyl residue that is donated by folic acid (folate) (Fig. 1). Numerous nutraceuticals directly supply methyl donors in the methyl activation cycle or serve as cofactors in this metabolic pathway which includes methionine, vitamin B_6, betaines (trimethylglycine), vitamin B_{12}, and zinc. In this chapter, I would like to review the relation between fetal nutritional environment and epigenetic modification as predisposing factors, the metabolism of one-carbon metabolism, epigenetic change of PPARα in malnutrition in utero, and fatty acid desaturases induced by PPARα.

2 Importance of Folic Acid in One-Carbon Metabolism

It has been known that deficiency dietary of folic acid in the periconceptional and early stage of pregnancy causes serious congenital anomalies, such as neural tube defect or other central nervous system defects. It is recommended globally for women to take folic acid and multivitamins from 1 to 3 months prior to conception throughout the pregnancy. In fact, folic acid is now fortified to grain or cereal in North and South America.

Folic acid is an important nutrient that acts as donor for methyl residue that is a potent regulator in epigenetic gene regulation. One carbon metabolism is the metabolic pathway that consists of methyl residue donating cycle, nucleotide providing and sulfur transsulfuration (Fig. 1). Many enzymes and coenzymes, such as vitamins B_2, B_6, and B_{12}, and zinc, participate to this system. Final metabolic product, S-adenosylmethionine (SAM), donates the methyl residue to the DNA, histone, and neurotransmitters. SAM is changed into S-adenosylhomocysteine (SAH) by releasing methyl residue, and then SAH is changed into homocysteine.

Homocysteine is methylated and changed into methionine via two enzymes: 5-methyltetrahydro folate-homocysteine methyltransferase (MTR) and betaine-homocysteine methyltransferase (BHMT), respectively [2]. MTR is an enzyme that catalyzes the final step of methionine biosynthesis from 5-methyl tetrahydrofolate (5-methyl-THF) and homocysteine. It is known as methionine synthase. MTR contains methylcobalamin (Me-Vitamin B_{12}) as cofactor. 5-Methyl-THF and homocysteine are used as substrates. This enzymatic reaction proceeds with a two-step ping-pong mechanism. At the first step, MTR transfers methyl residue from 5-methyl-THF to cobalamin (vitamin B_{12}; VB_{12}), and Me-VB_{12} and tetrahydrofolic acid (THF) are produced. At the second step, methyl residue is transferred from Me-B_{12} to homocysteine, and methionine is produced, and VB_{12} is regenerated (Fig. 1). Betaine-homocysteine S-methyltransferase (BHMT) is expressed most predominantly in the liver and kidney. BHMT transfers methyl residue from betaine (trimethylglycine) and a hydrogen ion from homocysteine to produce dimethylglycine and methionine, respectively.

In the sulfur transfer cycle, cystathionine-β-synthase (CBS) converts homocysteine to cystathionine by transsulfuration. In this cycle, cystathionine γ lyase converts to cysteine using vitamin B_6 as a coenzyme (Fig. 1). CBS occupies a pivotal position in sulfur metabolism at the homocysteine junction to convert cystathionine to methionine or convert cystathionine to cysteine through the transsulfuration pathway. Moreover, the transsulfuration pathway is the only one that is capable of removing excess sulfur-containing amino acids [3].

In addition to the preventive effect of malformations, folic acid functions as a donor of the methyl residue that contributes· to epigenetic modification. Proper intake of supplements or food is important from periconceptional period throughout pregnancy. Undesirable epigenetic modifications can be caused by higher or lower amount of folic acid. With food fortification of folic acid, unmetabolized folic acid (UMFA) was recently detected in blood samples collected in the US and Canada surveys [4]. The detection of UMFA in the blood indicates saturation of dihydrofolate reductase in the liver and also suggests excess intake of folic acid, which potentially cause cognitive dysfunction or cancer in adult by changing epigenetic modification [5]. Therefore, the serum UMFA level can be a marker of high intake of folic acid. Based on this concept, upper limit of folic acid intake has recently been determined (i.e., 1000 μg per day) in Garman.

Since homocysteine produces radical oxygen and inhibit NO production, individuals with high homocysteine levels in blood have higher risk of thrombosis, stroke, Alzheimer's disease, and cardiovascular diseases [6]. Whereas the prevalence of mild hyperhomocysteinemia is only 5–7%, it is detected in 30% of the patients with coronary heart disease and in 42% of the patients with cerebrovascular disorders [7].

Blood level of homocysteine, as an intermediate metabolite, is thought to be another marker of one-carbon metabolism homeostasis. Either high or low level of homocysteine indicates abnormal metabolic state: high level of homocysteine indicates low intake of folic acid or VB_{12}, and low level indicates high intake of these nutrients. Administration of folic acid or VB_{12} to the patients with peripheral circulatory insufficiency with high homocysteine level improved various symptoms [8]. Therefore, the optimal amount intake of folic acid is important for pregnant women not only to prevent congenital anomalies (e.g. neural tube defects) but also to reduce the risk of various life style related diseases.

3 Abnormal Lipid Metabolism (Dyslipidemia) Affected by Fetal Malnutrition

Dyslipidemia is an important risk factor of coronary heart disease, which is one of the leading causes of death in developing and developed counties. Studies on Dutch winter famine showed that prenatal malnutrition was associated with the dyslipidemia (e.g., elevated total cholesterol and triglycerides) in females [9]. Studies on Jewish holocaust survivors and Chinese famine also revealed that early-life famine caused dyslipidemia in males and females [10, 11].

Lillycrop and Burdge et al. investigated epigenetic change of the genes associated with lipid metabolism in weaning rats [12]. Protein-restricted diet to the pregnant rat decreased DNA methylation levels of peroxisomal proliferator-activated receptor α (*Pparα*) by 23% and glucocorticoid receptor (*Gr*) by 21% in rat offspring after weaning and increased expression of these genes, respectively. Expression of acyl-CoA oxidase 1 (*Acox1*) was also increased. On the contrary, protein-restricted diet did not change methylation status and expression of the *Pparγ* gene. These epigenetic abnormalities were observed not only on day 34 but also on day 80 after birth, suggesting that perinatal malnutrition induces persistent epigenetic changes in the genes associated with lipid metabolism.

PPARα is a key regulatory factor in lipid metabolism, and its epigenetic changes induced overexpression of fatty acid desaturase 1 (also designated delta-5 desaturase), fatty acid desaturase 2 (also designated delta-6 desaturase), and *Acox1* genes, which subsequently facilitate β oxidation [12]. Delta-5 desaturase and delta-6 desaturase are involved in polyunsaturated fatty acid (PUFA) metabolism and are thought to be associated with the risk of incident metabolic syndrome [13]. Expression of delta-6 desaturase, a rate-limiting enzyme in this metabolic pathway, is enhanced by PPARα.

Another study was conducted in which either WY14643 (PPARα agonist), safflower oil containing linoleic acid (18:2 (n-6)), Menkladen fish oil containing eicosapentaenoic acid (20:5 (n-3)) and docosahexaenoic acid (22:6 (n-3)), or triolein oil containing oleic acid (18:1 (n-9)) used as control was given to pregnant rats for 5 days (Fig. 2). As a result, WY14643 upregulated delta-6 desaturase whereas PUFA (e.g., fish oil and safflower oil) downregulated delta-6 desaturase. The underlying mechanism of differential expression is an upstream responsive element for PPARα and a downstream negative responsive element for PUFA within the 5′-flanking region of the *Fads 2* gene that encodes delta-6 desaturase [14].

Taken together, malnutrition in utero causes an epigenetic change in the PPARα gene, which accelerates the activity of delta-6 desaturase, resulting in incident metabolic syndrome including dyslipidemia in adulthood [15].

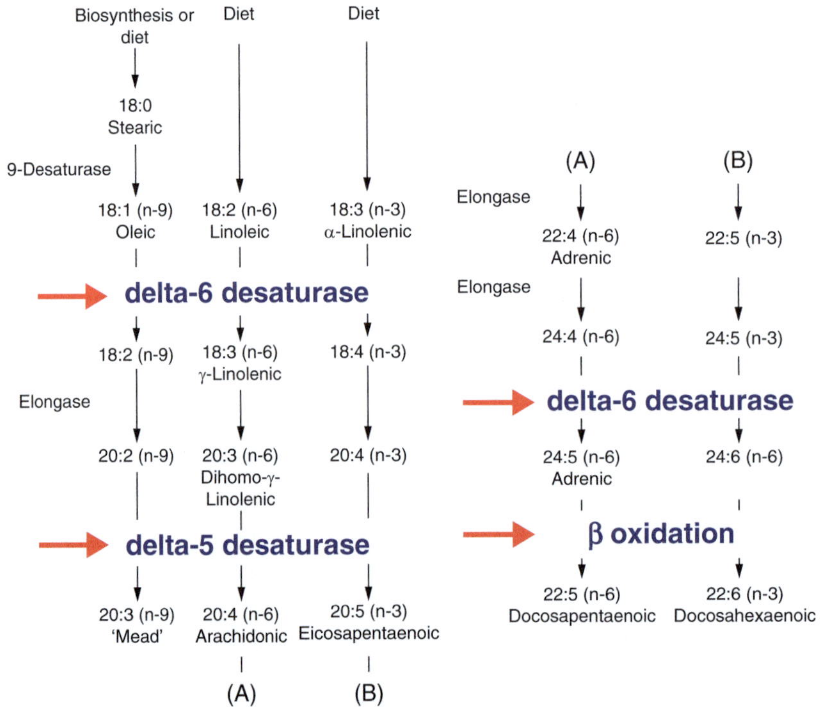

Fig. 2 Metabolic pathway of polysaturated fatty acids **Delta-5 desaturase (Fads 1) and delta-6 desaturase (Fads 2) desaturate polyunsaturated fatty acids. Particularly, Fads 2 induces desaturation of linoleic acid (18:2 (n-6)) and linolenic acid (18:3 (n-3)) and produces arachidonic acid (20:4 (n-6)) and docosahexaenoic acid (22:6 (n-3)), respectively**

4 Dyslipidemia of the Offspring Induced by High-Fat Diet During Pregnancy

Several lines of evidence suggest that intake of a large amount of specific types of lipids in the early life induces divergent metabolic changes especially in vessels. For example, administration of a large amount of saturated fatty acid to pregnant animals causes cardiometabolic features in their offspring, which include hypertension, arteriosclerosis, decreased number of endothelial cells and vascular smooth muscle cells, and decreased activity of the aortic Na^+/K^+ ATPase [16]. In addition, excess intake of trans-fatty acids in infancy decreases serum levels of the 20:4 (n-6) fatty acid and the 22:6 (n-3) fatty acid, which increases the risk of cardiovascular diseases in later life [17].

Furthermore, deficiency of n-3 PUFAs during pregnancy and lactation cause hypertension in adult offspring [18]. In spontaneous hypertensive SHR rats, production of arachidonic acid (20: 4 (n-6)) is elevated in the vascular endothelium accompany, and this increases serum level of 20:4 (n-6), which causes hypertension. The 22:6 (n-3) and 20:4 (n-6), fatty acids which play important roles in the brain and cardiovascular system, are taken from foods and are also synthesized in vivo from the precursors such as linolenic acid (18:3 (n-3)) and linoleic acid (18:2 (n-6)), respectively (Fig. 2) [18]. There are many kinds of elongating and unsaturation enzymes for long-chain fatty acids in the liver and the brain. A similar metabolic system also exists in the blood vessel cells. Genetic polymorphisms are known in the genes encoding delta-5 desaturase and delta-6 desaturase (*Fads 1* and *Fads 2*, respectively), and specific genotypes of these genes are known to be associated with increased risk of cardiovascular diseases [19].

Kelsall et al. reported that overdose administration of fatty acids to maternal pregnant rats altered epigenetic modification of the genes associated with lipid metabolism in the arterial endothelium of the offspring, which change the vascular relaxation and contracting function for a long time after birth [20]. As fat component in dosed nutrients, safflower oil containing 18:2 (n-6), hydrogenated soybean oil containing trans-fatty acid, butter containing saturated fatty acids, and Menkladen oil containing 22:6 (n-3) and 20:5 (n-3) were given to pregnant rats. Proportion of each fatty acid was 7% and 21% of the total energy. As a result, the relaxation ability of aortic vascular was dose-dependently decreased, and the constriction was dose-dependently increased on day 77 after delivery. Eicosanoids such as PGE 2, PGF 2alfa, and TBXA 2 are involved in this excessive vasoconstriction. These eicosanoids are formed from 20:4 (n-6) with cytosolic phospholipase A 2 and cyclooxygenase. Although the high-fat diet did not alter DNA methylation status in the promoter region of *Fads 1*, this diet increased DNA methylation in the promoter region of *Fads 2* [14], suggesting that Fads 2 expression may be downregulated with high-fat diet in the arterial smooth muscle.

Heijmans et al. investigated epigenetic status of genes in the peripheral mononuclear cells of the individuals who encountered Dutch hunger and were subjected to malnutrition in the fertilized period. They found decreased DNA methylation in CpG islands in the promoter region of the imprinted insulin-growth factor 2 (*IGF-2*) [21]. On the other hand, they found increased DNA methylation in imprinted maternally expressed 3 (*MEG 3*), *IL-10*, *LEPTIN*, ATP-binding cassette A1, and guanine nucleotide-binding protein genes. These findings seem to be paradoxical at first sight, but they may indicate importance of understanding epigenetics at genome-wide level, not single gene level. Precise mechanism for nutrition-induced differential methylation, which was shown in this study, remains to be elucidated.

5 Conclusion

In this chapter, I reviewed one-carbon metabolism and lipid metabolism which are underlying basic mechanisms of DOHaD from the genetic and epigenetic viewpoints. The nutritional environment from the periconception to the neonatal period can alter epigenetic modifications of genes that are associated with lipid metabolism. Either insufficient or excessive nutrition can change epigenetic modification that encodes essential enzymes for lipid metabolism. The altered state can further persist during one's lifetime, which potentially causes noncommunicable diseases in adulthood. Genetic studies revealed a precise mechanism of regulation of the delta-6 desaturase gene (*Fads 2*), which has a positive element and a negative responsive element for PPARα and PUFA, respectively, in the promoter region. Based on this context, abnormal nutrition-induced hypomethylation of the PPARα-binding positive element within the promoter region of *Fads 2* upregulated gene expression, which presumably induced delta-6 desaturase, and this may contribute to develop dyslipidemia in adulthood. Further studies will elucidate the molecular and epigenetic mechanism associated with nutritional factors for other features of incident metabolic syndrome such as hypertension and arteriosclerosis.

References

1. Hanson M, Godfrey KM, Lillycrop KA, Burdge GC, Gluckman PD. Developmental plasticity and developmental origins of non-communicable disease: theoretical considerations and epigenetic mechanisms. Prog Biophys Mol Biol. 2011;106:272–80.
2. Matthews RG, Smith AE, Zhou ZS, Taurog RE, Bandarian V, Evans JC, Ludwig M. Cobalamin-dependent and cobalamin-independent methionine synthases: are there two solutions to the same chemical problem? Helv Chim Acta. 2003;86:3939–54.
3. Janosík M, Kery V, Gaustadnes M, Maclean KN, Kraus JP. Regulation of human cystathionine beta-synthase by S-adenosyl-L-methionine: evidence for two catalytically active conformations involving an autoinhibitory domain in the C-terminal region. Biochemistry. 2001;40:10625–33.
4. Hu J, Wang B, Sahyoun NR. Application of the key events dose-response framework to folate metabolism. Crit Rev Food Sci Nutr. 2016;56:1325–33.
5. Wiens D, DeSoto MC. Is high folic acid intake a risk factor for autism? – a review. Brain Sci. 10 Nov 2017;7(11). pii: E149. https://doi.org/10.3390/brainsci7110149.
6. Lawrence de Koning AB, Werstuck GH, Zhou J, Austin RC. Hyperhomocysteinemia and its role in the development of atherosclerosis. Clin Biochem. 2003;36:431–41.
7. Clarke R, Daly L, Robinson K, Naughten E, Cahalane S, Fowler B, Graham I. Hyperhomocysteinemia: an independent risk factor for vascular disease. N Engl J Med. 1991;324:1149–55.
8. Zheng Z, Wang J, Yi L, Yu H, Kong L, Cui W, Chen H, Wang C. Correlation between behavioural and psychological symptoms of Alzheimer type dementia and plasma homocysteine concentration. Biomed Res Int. 2014;2014:383494.
9. Lumey LH, Stein AD, Kahn HS, Romijn JA. Lipid profiles in middle-aged men and women after famine exposure during gestation: the Dutch hunger winter families study. Am J Clin Nutr. 2009;89:1737–43.
10. Keinan-Boker L, Shasha-Lavsky H, Eilat-Zanani S, Edri-Shur A, Shasha SM. Chronic health conditions in Jewish holocaust survivors born during world war II. Isr Med Assoc J. 2015;17:206–12.
11. Wang Z, Li C, Yang Z, Ma J, Zou Z. Fetal and infant exposure to severe Chinese famine increases the risk of adult dyslipidemia: results from the China health and retirement longitudinal study. BMC Public Health. 2017;17:488.
12. Lillycrop KA, Phillips ES, Jackson AA, Hanson MA, Burdge GC. Dietary protein restriction of pregnant rats induces and folic acid supplementation prevents epigenetic modification of hepatic gene expression in the offspring. J Nutr. 2005;135(6):1382.
13. Yary T, Voutilainen S, Tuomainen TP, Ruusunen A, Nurmi T, Virtanen JK. Omega-6 polyunsaturated fatty acids, serum zinc, delta-5- and delta-6-desaturase activities and incident metabolic syndrome. J Hum Nutr Diet. 2017;30:506–14.
14. Tang C, Cho HP, Nakamura MT, Clarke SD. Regulation of human delta-6 desaturase gene transcription: identification of a functional direct repeat-1 element. J Lipid Res. 2003;44:686–95.
15. Burdge GC, Lillycrop KA. Nutrition, epigenetics, and developmental plasticity: implications for understanding human disease. Annu Rev Nutr. 2010;30:315–39.
16. Giannotti G, Landmesser U. Endothelial dysfunction as an early sign of atherosclerosis. Herz. 2007;32:568–72.

17. Decsi T. Koletzko BDo trans fatty acids impair linoleic acid metabolism in children? Ann Nutr Metab. 1995;39:36–41.
18. Taylor PD, Khan IY, Hanson MA, Poston L. Impaired EDHF-mediated vasodilatation in adult offspring of rats exposed to a fat-rich diet in pregnancy. J Physiol. 2004;558:943–51.
19. Simopoulos AP. Genetic variants in the metabolism of omega-6 and omega-3 fatty acids: their role in the determination of nutritional requirements and chronic disease risk. Exp Biol Med. 2010;235:785–95.
20. Kelsall CJ, Hoile SP, Irvine NA, Masoodi M, Torrens C, Lillycrop KA, Calder PC, Clough GF, Hanson MA, Burdge GC. Vascular dysfunction induced in offspring by maternal dietary fat involves altered arterial polyunsaturated fatty acid biosynthesis. PLoS One. 2012;7:e34492.
21. Heijmans BT, Tobi EW, Stein AD, Putter H, Blauw GJ, Susser ES, Slagboom PE, Lumey LH. Persistent epigenetic differences associated with prenatal exposure to famine in humans. Proc Natl Acad Sci USA. 2008;105:17046–9.

Novel Models of Epigenetic Gene Regulation in the Nutritional Environment

Kazuki Mochizuki, Natsuyo Hariya, and Takeo Kubota

Abstract

Epigenetic memories are acquired information included in the chromatin or DNA such as methylation and histone modifications. Recent studies suggest that epigenetic memories determine the types of differentiated cells in each tissue. Moreover, the development of metabolic diseases induced by environmental factors during development is controlled by epigenetic regulation rather than the genetic regulation such as DNA sequence-dependent transcriptional regulation. In general, the demethylation of CpG islands induces histone acetylation, associated changes from heterochromatin to euchromatin, and enhances transcriptional activation. Under the classical model of epigenetics, these changes are induced by the binding of transcriptional factors to cis-elements located on promoter/enhancer regions and the associated binding of histone acetyl-transferase and the transcription initiation complex. This model is dependent on epigenetics in the promoter/enhancer region and is used to explain the induction of genes by lipophilic nutrients such as vitamin A, vitamin D, and unsaturated fatty acid metabolites. However, recent studies have demonstrated that epigenetics in the gene body (transcribed region) also regulate transcription. This novel model postulates that histone acetylation and bromodomain-containing protein 4, which contains two bromodomains to bind acetylated histones, on the gene body enhance transcriptional elongation. Gene expression alterations induced by carbohydrate signals and changes to energy balance in the body accompanied by the intake of major nutrients are also regulated by this model. In this section, we introduce these epigenetic regulations and their relationship with nutrient intake and discuss the link between epigenetic regulation and the development of metabolic diseases.

K. Mochizuki (✉)
Faculty of Life and Environmental Sciences, University of Yamanashi, kofu-shi, Yamanashi, Japan
e-mail: mochizukik@yamanashi.ac.jp

N. Hariya
Faculty of Health and Nutrition, Department of Nutrition, Yamanashi Gakuin University, kofu-shi, Yamanashi, Japan
e-mail: n-hariya@ygu.ac.jp

T. Kubota
Faculty of Child Studies, Seitoku University, Matsudo, Chiba, Japan

© Springer Nature Singapore Pte Ltd. 2018

T. Kubota, H. Fukuoka (eds.), *Developmental Origins of Health and Disease (DOHaD)*, Advances in Experimental Medicine and Biology 1012, https://doi.org/10.1007/978-981-10-5526-3_2

Keywords
Epigenetics · Transcription · Chromatin · DNA methylation · Histone modification

1 Introduction

The prevalence of metabolic diseases such as obesity and type 2 diabetes has been increased worldwide with changes in lifestyles including dietary habits. One example of these changes is the excessive dieting undertaken by Japanese women before and during pregnancy. The resulting environmental stresses experienced during fetal growth are considered to increase the risks of developing metabolic diseases such as obesity, type 2 diabetes, hypertension, and psychiatric diseases during later life (i.e., the developmental origins of health and disease (DOHaD) theory). Disease development is thought to be caused by the long-term exposure of fetal cells and tissues to environmental factors such as hypernutrition or malnutrition.

Several studies have suggested that epigenetic memory determines the type of differentiated cells in each tissue. Specifically, the induction of metabolic diseases by environmental factors is controlled by epigenetic rather than genetic regulation because these diseases develop from stored memories of environmental stimulation during developmental stages.

Here, we introduce epigenetic regulation and describe its relationship with metabolic diseases.

2 Basic Epigenetics

Transcriptional regulation is based on DNA sequences, in particular the promoter and enhancer regions upstream of the transcription initiation site. Transcription is activated by the binding of transcription factors to transcription factor binding sites within promoters and enhancers. This binding to cis-elements recruits the transcription initiation complex comprising RNA polymerase II to bind the TATA box and initiate transcription (Fig. 1a). Thus, the regulation is dependent on the genetic information of cis-elements but can be theoretically switched off by disassociating transcriptional factors.

Recent studies suggest that transcriptional regulation is also controlled by epigenetic memories, such as the methylation of CpG islands in the promoter region of genes. DNA methylation involves the addition of a methyl group to the cytosine of DNA. Cytosine residues in CG repeat sequences are typically methylated and are known as CpG islands. When CpG islands are hypermethylated, the chromatin surrounding the gene changes from euchromatin to heterochromatin by associating with methylated DNA-binding proteins, thus repressing transcription. Conversely, demethylation of CpG islands changes heterochromatin to euchromatin, thus inducing transcription through histone modifications such as histone acetylation (Fig. 1b).

Other histone modifications include methylation, phosphorylation, and ubiquitination (Fig. 2). Within chromatin, histones interact with DNA to form nucleosomes, which consist of histone octamers (two molecules each of H2A, H2B, H3, and H4). The methylation of histone H3 at lysine 4 (K4) is generally related to the activation of transcription, whereas methylation of histone H3K9/K27 and H4K20 is associated with transcriptional repression. Histone acetylation relaxes the chromatin by changing its charge and recruits proteins related to transcription. Histone methylation does not alter the chromatin charge but enables changes to occur between heterochromatin and euchromatin and transcription activation/repression through recruiting various proteins to the chromatin [1].

Heterochromatin formation occurs through the binding of DNA methyltransferase (DNMT) to heterochromatin protein (HP) on methylated histone H3K9, which catalyzes the methylation of CpG islands. The demethylation of CpG islands and subsequent histone acetylation via histone acetyl-transferase (HAT) induce euchromatin formation by removing the HP (Fig. 3).

Epigenetic memories are thought to be sustained over long periods. Indeed, the induction of histone H3K4 mono-methylation around the inflammatory gene nuclear factor-κB by high

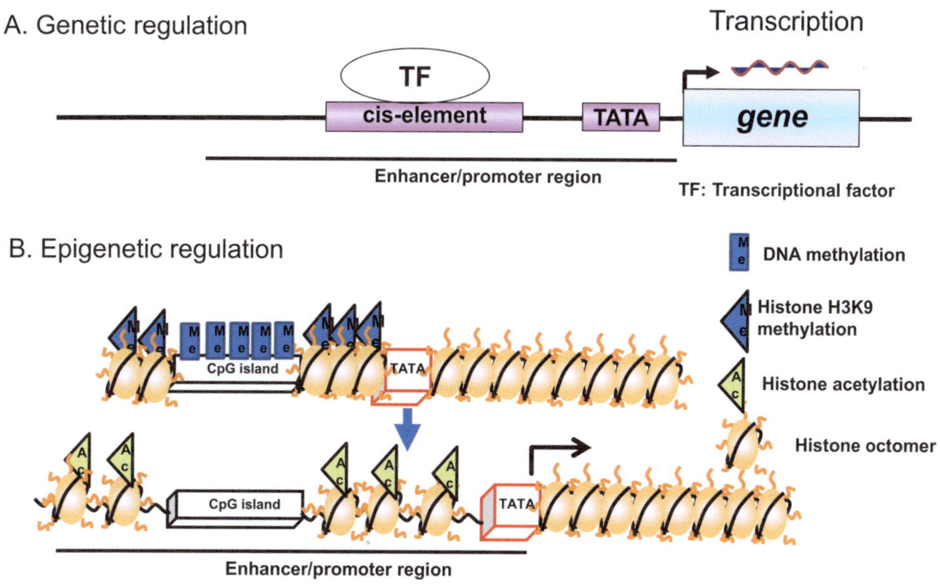

Fig. 1 Genetic and epigenetic regulation. (**a**) Genetic regulation. (**b**) Epigenetic regulation

Fig. 2 Histone modifications. (**a**) Histone H2A. (**b**) Histone H2B. (**c**) Histone H3. (**d**) Histone H4

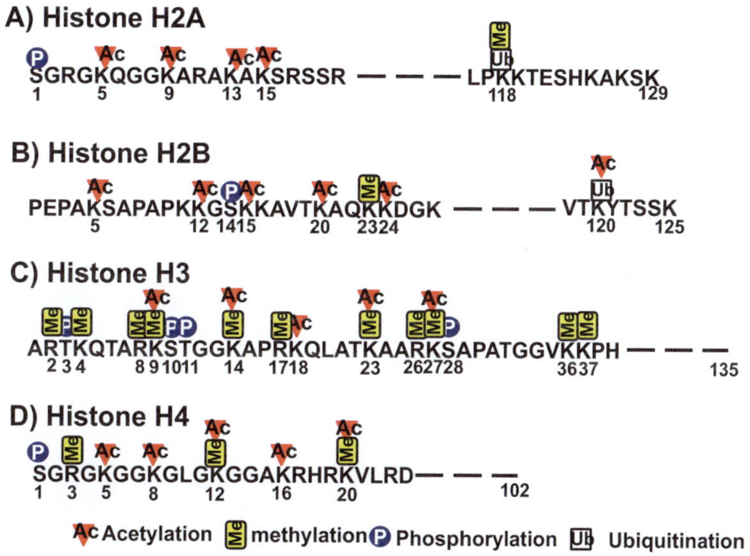

glucose exposure of primary human aortic endothelial cells for 16 h was maintained for 6 days when the cells were incubated under normal glucose conditions [2].

The importance of epigenetic memories for tissue development and the progression of metabolic diseases has been shown by studies using gene targeting methods in mice. Heterozygous gene targeting of a HAT CREB-binding protein reduced the development of high-fat diet-inducible impaired glucose tolerance [3]. Mice that were heterozygous for bromodomain-containing protein 4 (BRD4), which binds acetylated histones, showed lower levels of adiposity compared with wild-type mice [4]. On the other hand, BRD2 disruption led to adiposity [5]. Mice with a hepatic deficiency of the histone deacetylase sirtuin 1 and mice lacking the gene expres-

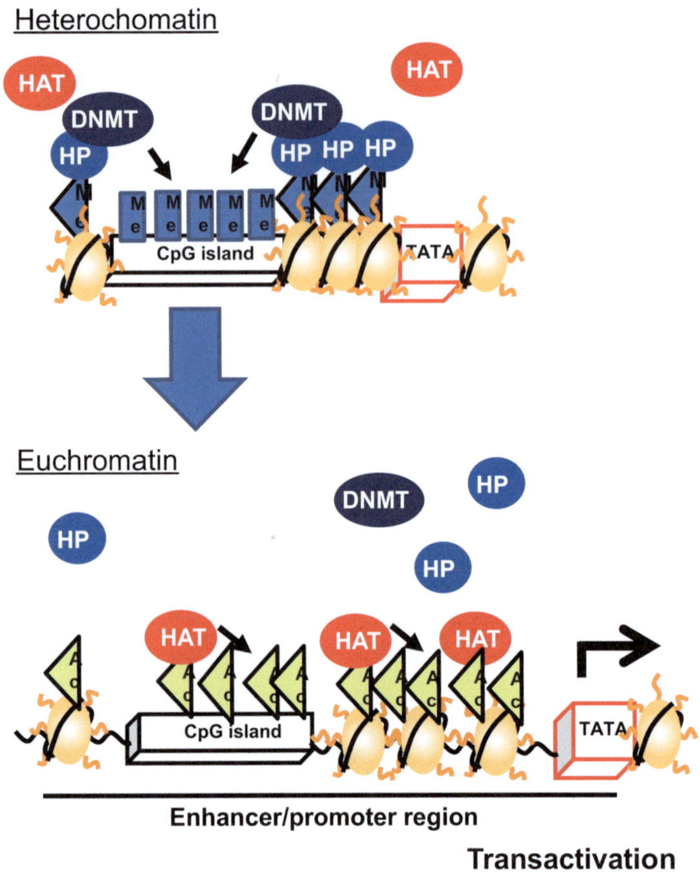

Fig. 3 Changes from heterochromatin to euchromatin

sion of JmjC-containing H3K9 demethylase 2a, a demethylase of histone H3K9, were obese and demonstrated insulin resistance and lipid abnormalities compared with wild-type mice [6, 7]. Similarly, transgenic mice overexpressing the DNA methyltransferase *Dnmt3a* became obese and developed inflammation in their adipose tissue [8]. These results suggest that the disruption of epigenetic regulation is related to the development of metabolic diseases such as obesity and type 2 diabetes. Therefore, epigenetic regulation is involved not only in the determination of cell differentiation but also in the onset of chronic diseases through lifestyle accumulation. Additionally, studies of targeted disruption of genes during fetal stages revealed that epigenetic memories are important for fetal development and the development of diseases under the DOHaD theory as well as in adulthood. Epigenetic disruption observed in animals that underwent environmental stresses such as malnutrition during various developmental stages is described in the next section.

3 Classical and Novel Epigenetics

It is not fully understood how nutrient signals during different developmental stages affect metabolism and associated metabolic abnormalities in later life. This section introduces the response of classical and novel epigenetics to nutrient signals.

Under the classical model, transcriptional factor-associated HAT acetylates histones in promoter and enhancer regions, which recruit transcriptional initiation complexes with bromodomain

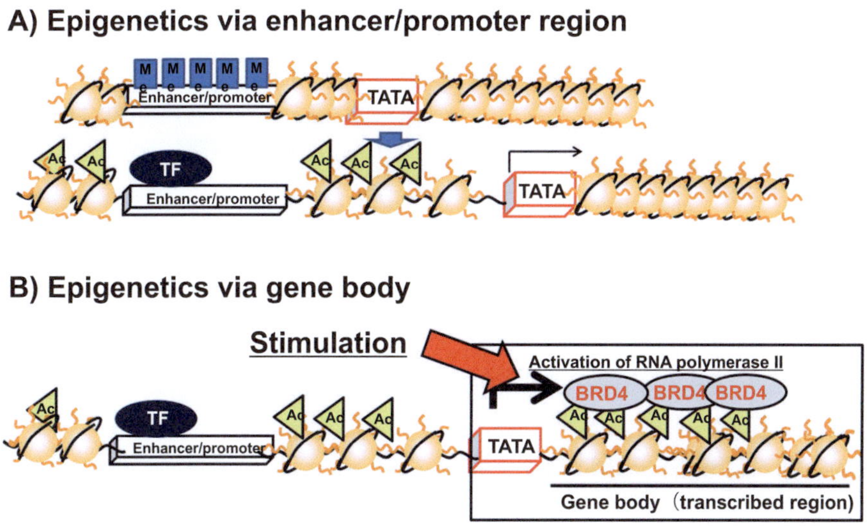

Fig. 4 Epigenetics via the enhancer/promoter region and via the gene body. (**a**) Classic epigenetics. (**b**) Novel epigenetics

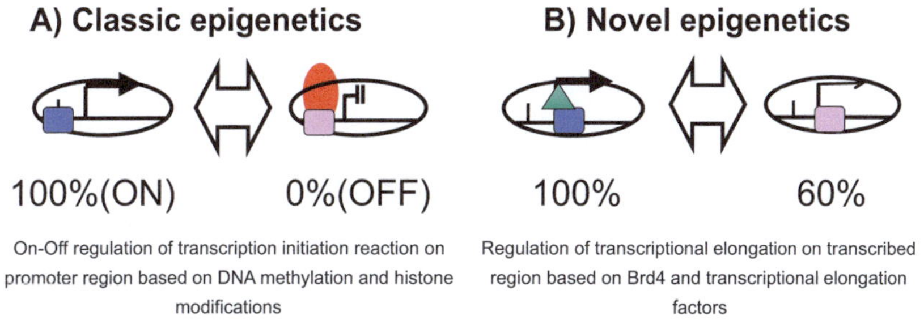

Fig. 5 Differences between classic and novel epigenetic models. (**a**) Epigenetics via the promoter/enhancer region. (**b**) Epigenetics via the gene body

proteins. These bind acetylated histones around the promoter and enhancer regions and initiate transcription (Fig. 4a). The regulation is conceptually dependent on transcriptional factors binding to cis-elements. Nutrient signals, such as those from vitamin A, vitamin D, and unsaturated fatty acid metabolites, are transmitted to the nuclear receptors retinoic acid receptor, vitamin D receptor, and peroxisome proliferator-activated receptor, respectively, and are dependent on the regulation [9]. The classical epigenetics model describes an on–off regulation of transcription initiation by epigenetic memories such as DNA methylation and histone modifications. The model regulates transcriptional changes such as from 0% to 100% or 100% to 0% (Fig. 5a).

Recent studies have demonstrated that histone acetylation around the gene body (transcribed region) is important for transcriptional control. The acetylation of histone and associated BRD4 in the gene body enhances transcriptional elongation (Fig. 4b), which controls RNA polymerase II activity. Enhanced acetylation of histones in the gene body recruits BRD4 and positive elongation factor b (P-TEFb) (a cyclin T1-CDK9 complex). After forming the transcriptional initiation complex with RNA polymerase II on the TATA box, general transcription factor

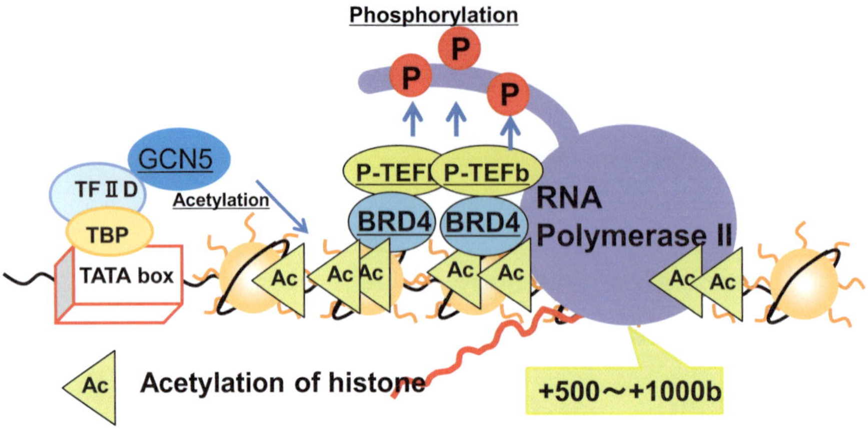

Fig. 6 Acetylated histone-BRD4-P-TEFb enhances the transcriptional elongation complex

IIH initiates transcriptional elongation by phosphorylating the fifth serine residue at the C-terminal domain (CTD) of RNA polymerase II. BRD4 enhances the slide of RNA polymerase II from +500 to +5000 bp along the gene body from the transcription initiation site. It also binds acetylated histones in the gene body and recruits the P-TEFb complex which phosphorylates RNA polymerase II at the second serine residue of the CTD, resulting in its activation and enhancing transcriptional elongation [10, 11] (Fig. 6).

Under the novel model, epigenetic modifications regulate the efficacy of mRNA synthesis, for example, from 60% to 100% (Fig. 5b). BRD4 enhances the transcriptional elongation reaction of genes related to cell cycle progress. For example, during the G1–S transition, BRD4 activates RNA polymerase II on the gene body of cell cycle-related genes such as *Cyclin D1*, *Orc2*, and *Mcm2* [12]. The expression of such genes is also regulated by transcriptional factors such as E2F. We previously demonstrated that the induction of genes by glucocorticoids is associated not only with binding of the glucocorticoid receptor to cis-elements but also with histone acetylation and BRD4-P-TEFb binding around the gene body of the glucocorticoid-responsive gene *Glut5* in small intestinal cell lines [13, 14]. These results indicate that transcriptional elongation is necessary for many aspects including cis-element-dependent mRNA expression.

4 The Association of Epigenetic Gene Regulation with Nutrients

Transcriptional elongation is associated with the expression changes of genes induced by nutritional intake and the related development of metabolic diseases such as obesity, type 2 diabetes, and cardiovascular diseases. This reflects the fact that the intake of major nutrients controls the level of mRNA for metabolic genes. However, it remains unclear whether acetylated histones and BRD4 in the gene body regulate nutrient signal-inducible gene expression and the development of lifestyle-related diseases. We recently demonstrated that high carbohydrate intake enhances histone acetylation in the gene body, albeit less so in the promoter/enhancer region, of carbohydrate-responsive genes. We also showed that acetylated histones and/or BRD4 regulate carbohydrate digestion-/absorption-related genes such as sucrase-isomaltase, *Sglt1*, and *Glut5* [15, 16]. Fat accumulation-related genes in the liver, such as *Cyp8b1*, *Dak*, and *Plin5*, are similarly regulated by BRD4 [17]. Additionally, we found that histone acetylation in the gene body is related to adiponectin gene induction in 3T3-L1 adipocytes [18], while the induction of fatty acid synthase (*Fas*) in the liver during the suckling–weaning transition period is associated with histone acetylation in the *Fas* gene body [19]. Specifically, the expression of genes induced by carbohydrate intake appears to be associated with histone acetylation in the gene body rather than the binding of transcriptional factors to cis-elements. This indicates that carbohydrate-responsive genes and many genes responsive to the energy balance are regulated by BRD4.

Although it is not certain whether the epigenetic information in the gene body regulates the development of metabolic diseases, the induction of *Fas* in a metabolic syndrome rat model was shown to be associated with histone acetylation in the *Fas* gene body [20]. Additionally, the BRD4 inhibitor (+)-JQ1 was reported to reduce the expression of genes related to the development of fatty liver, such as *Cyp8b1*, *Dak*, and *Plin5* [17]. *Brd4* heterozygous mice also showed a lower weight of adipose tissue compared with wild type [4]. Furthermore, many studies, including our own, demonstrated that the development of metabolic diseases in rodent models is associated with an increase (e.g., fat accumulation-related genes in the liver) or decrease (e.g., insulin-sensitive genes such as adiponectin) of target genes of BRD4 and/or acetylated histones in the gene body.

Taken together, BRD4 and epigenetics in the gene body appear to be involved with the expression of genes induced by major nutrient signals and by changes of the energy balance, as well as the development of metabolic disorders. Further studies should be conducted to verify this hypothesis.

5 Conclusion

Two types of epigenetic model are known. The first is the classical model, involving the transcriptional initiation reaction regulated by nuclear transcriptional factors on the cis-element of the promoter/enhancer region and associated histone modifications. The induction of genes by lipophilic nutrients such as vitamin A and vitamin D is regulated by this model. The second novel model involves the transcriptional elongation reaction regulated by histone modifications in the gene body. Major nutrient signals, in particular carbohydrate signals, and those of the energy balance in the body, are regulated in this model. However, it is still necessary to clarify the nutrient signals that are transmitted to the histone modifications in the gene body, the links between classical and novel epigenetic models, and the best model that can explain epigenetic regulation associated with the development of metabolic diseases for better understanding of DOHaD.

References

1. Bruce Alberts AJ, Lewis J, Morgan D, Raff M, Roberts K, Walter P. Molecular biology of the cell. 6th ed., paper ed. New York: Garland Science; 2015.
2. El-Osta A, Brasacchio D, Yao D, Pocai A, Jones PL, Roeder RG, Cooper ME, Brownlee M. Transient high glucose causes persistent epigenetic changes and altered gene expression during subsequent normoglycemia. J Exp Med. 2008;205:2409–17.
3. Yamauchi T, Oike Y, Kamon J, Waki H, Komeda K, Tsuchida A, Date Y, Li MX, Miki H, Akanuma Y, Nagai R, Kimura S, Saheki T, Nakazato M, Naitoh T, Yamamura K, Kadowaki T. Increased insulin sensitivity despite lipodystrophy in Crebbp heterozygous mice. Nat Genet. 2002;30:221–6.
4. Houzelstein D, Bullock SL, Lynch DE, Grigorieva EF, Wilson VA, Beddington RS. Growth and early postimplantation defects in mice deficient for the bromodomain-containing protein Brd4. Mol Cell Biol. 2002;22:3794–802.
5. Wang F, Liu H, Blanton WP, Belkina A, Lebrasseur NK, Denis GV. Brd2 disruption in mice causes severe obesity without type 2 diabetes. Biochem J. 2009;425:71–83.
6. Wang RH, Kim HS, Xiao C, Xu X, Gavrilova O, Deng CX. Hepatic Sirt1 deficiency in mice impairs mTorc2/Akt signaling and results in hyperglycemia, oxidative damage, and insulin resistance. J Clin Invest. 2011;121:4477–90.
7. Inagaki T, Tachibana M, Magoori K, Kudo H, Tanaka T, Okamura M, Naito M, Kodama T, Shinkai Y, Sakai J. Obesity and metabolic syndrome in histone demethylase JHDM2a-deficient mice. Genes Cells. 2009;14:991–1001.
8. Kamei Y, Suganami T, Ehara T, Kanai S, Hayashi K, Yamamoto Y, Miura S, Ezaki O, Okano M, Ogawa Y. Increased expression of DNA methyltransferase 3a in obese adipose tissue: studies with transgenic mice. Obesity (Silver Spring). 2010;18:314–21.
9. Xu L, Glass CK, Rosenfeld MG. Coactivator and corepressor complexes in nuclear receptor function. Curr Opin Genet Dev. 1999;9:140–7.
10. Jang MK, Mochizuki K, Zhou M, Jeong HS, Brady JN, Ozato K. The bromodomain protein Brd4 is a positive regulatory component of P-TEFb and stimulates RNA polymerase II-dependent transcription. Mol Cell. 2005;19:523–34.

11. Hirose Y, Ohkuma Y. Phosphorylation of the C-terminal domain of RNA polymerase II plays central roles in the integrated events of eucaryotic gene expression. J Biochem. 2007;141:601–8.
12. Mochizuki K, Nishiyama A, Jang MK, Dey A, Ghosh A, Tamura T, Natsume H, Yao H, Ozato K. The bromodomain protein Brd4 stimulates G1 gene transcription and promotes progression to S phase. J Biol Chem. 2008;283:9040–8.
13. Takabe S, Mochizuki K, Goda T. De-phosphorylation of GR at Ser203 in nuclei associates with GR nuclear translocation and GLUT5 gene expression in Caco-2 cells. Arch Biochem Biophys. 2008;475:1–6.
14. Inamochi Y, Nishiyama A, Kubota T, Ozato K, Goda T, Mochizuki K. Transcription elongation factor Brd4-P-TEFb accelerates intestinal differentiation-associated SLC2A5 gene expression. Biochem Biophys Rep. 2016;7:150–6.
15. Inoue S, Mochizuki K, Goda T. Jejunal induction of SI and SGLT1 genes in rats by high-starch/low-fat diet is associated with histone acetylation and binding of GCN5 on the genes. J Nutr Sci Vitaminol (Tokyo). 2011;57:162–9.
16. Honma K, Mochizuki K, Goda T. Induction by fructose force-feeding of histone H3 and H4 acetylation at their lysine residues around the Slc2a5 gene and its expression in mice. Biosci Biotechnol Biochem. 2013;77:2188–91.
17. Yamada A, Honma K, Mochizuki K, Goda T. BRD4 regulates fructose-inducible lipid accumulation-related genes in the mouse liver. Metabolism. 2016;65:1478–88.
18. Sakurai N, Mochizuki K, Goda T. Modifications of histone H3 at lysine 9 on the adiponectin gene in 3T3-L1 adipocytes. J Nutr Sci Vitaminol (Tokyo). 2009;55:131–8.
19. Morishita S, Mochizuki K, Goda T. Bindings of ChREBP and SREBP1, and histone acetylation around the rat liver fatty acid synthase gene are associated with induction of the gene during the suckling-weaning transition. J Nutr Sci Vitaminol (Tokyo). 2014;60:94–100.
20. Suzuki T, Muramatsu T, Morioka K, Goda T, Mochizuki K. ChREBP binding and histone modifications modulate hepatic expression of the Fasn gene in a metabolic syndrome rat model. Nutrition. 2015;31:877–83.

Epigenetic Switching and Neonatal Nutritional Environment

Koshi Hashimoto and Yoshihiro Ogawa

Abstract

The hepatic metabolic function changes sequentially during early life in mammals to adapt to the drastic changes in the nutritional environment. Accordingly, hepatic fatty acid β-oxidation is activated after birth to produce energy from breast milk lipids. De novo lipogenesis is induced upon the onset of oral intake, when the major nutritional source switches to carbohydrate. However, how a particular metabolic pathway is activated during the liver maturation is poorly understood. We found that the expression of glycerol-3-phosphate acyltransferase 1 (GPAT1), a rate-limiting enzyme of de novo hepatic lipogenesis, is epigenetically regulated in the mouse liver by DNA methylation. In the neonatal liver, DNA methylation of the GPAT1 gene (*Gpam*) promoter, which is likely to be induced by DNA methyltransferase (Dnmt) 3b, inhibited the recruitment of sterol regulatory element-binding protein-1c (SREBP-1c), whereas in the adult, decreased DNA methylation resulted in active chromatin conformation, allowing the recruitment of SREBP-1c. Maternal nutritional environment affects the DNA methylation status in the *Gpam* promoter, GPAT1 expression, and triglyceride content in the liver of the offspring. We also found DNA demethylation and increased mRNA expression of the fatty acid β-oxidation genes in the postnatal mouse liver. The DNA demethylation is specifically induced in the lactation period. Analysis of mice deficient in the nuclear receptor peroxisome proliferator-activated receptor α (PPARα) and maternal administration of a PPARα ligand during the gestation and lactation periods reveals that the DNA demethylation is PPARα-dependent. These findings indicate the gene- and lifestage-specific DNA demethylation of a particular metabolic pathway in the neonatal liver to adapt the marked changes in nutritional environment in early life.

Keywords

DNA demethylation · De novo lipogenesis · GPAT1 · SREBP-1c · Dnmt3b · Fatty acid β-oxidation · PPARα

K. Hashimoto (✉)
Department of Preemptive Medicine and Metabolism, Tokyo Medical and Dental University, Tokyo, Japan
e-mail: khashimoto.mem@tmd.ac.jp

Y. Ogawa
Department of Molecular and Cellular Metabolism, Graduate School of Medical and Dental Sciences, Tokyo Medical and Dental University, Tokyo, Japan

Department of Medicine and Bioregulatory Science, Graduate School of Medical Sciences, Kyushu University, Fukuoka, Japan

AMED, CREST, Tokyo, Japan

© Springer Nature Singapore Pte Ltd. 2018
T. Kubota, H. Fukuoka (eds.), *Developmental Origins of Health and Disease (DOHaD)*, Advances in Experimental Medicine and Biology 1012, https://doi.org/10.1007/978-981-10-5526-3_3

1 Introduction

In the fetal period, the fetus is supplied by cord blood, which is glucose rich. After birth, upon the onset of breast-feeding, the major nutrition source should be changed to lipid derived from breast or formula milk. Then, after weaning, when oral intake begins, carbohydrate, which is glucose again, should be the major nutritional source [1] (Fig. 1). To adapt these drastic changes in the nutritional environment in the perinatal period, the metabolic function of the liver sequentially changes, so that it becomes matured as a major metabolic organ [1, 2]. For instance, after birth, fatty acid β-oxidation is activated to exchange milk lipids to energy in lactation period, and de novo hepatic lipogenesis is developed to produce lipid after weaning in the liver. However, how a particular metabolic pathway is activated during the liver maturation has been poorly understood.

2 Epigenetic Regulation of De Novo Triglyceride Biosynthesis

2.1 Epigenetic Regulation of Glycerol-3-Phosphate Acyltransferase 1 (GPAT1) Gene (*Gpam*)

Triglyceride (TG) is the major storage form of energy in animals. TG biosynthesis begins with the acylation of glycerol-3-phosphate by glycerol-3-phosphate acyltransferase 1 (GPAT1; *Gpam*) to form lysophosphatidic acid; this is the rate-limiting step in the hepatic TG biosynthesis pathway [3] (Fig. 2). In the acylation process, fatty acids produced by stearoyl-CoA desaturase 1 (SCD1; *Scd1*) and fatty acid synthase (FAS; *Fasn*) are used as acyl donors. Among the lipogenic enzymes, GPAT1 plays an important role in the regulation of hepatic TG biosynthesis [4, 5]. The lipogenic genes such as *Gpam, Scd1*, and *Fasn* are activated by sterol regulatory element-binding protein-1c (SREBP-1c), which is a transcription factor and a master regulator of lipogenesis. Indeed, their promoter regions contain the sterol regulatory elements (SREs) [6–8].

The GPAT1 mRNA is only slightly expressed in the neonatal liver but markedly increased in the adult liver. Similar to GPAT1, both SCD1 and FAS mRNA levels, as well as protein levels of GPAT1, SCD1, and FAS, and GPAT1-mediated TG/diacylglycerol (DG) biosynthesis are increased in the adult liver relative to the neonatal liver.

2.2 DNA Methylation of the *Gpam* Promoter

Epigenetics such as DNA methylation and histone modification is involved in the regulation of a diverse range of biological processes in mammals. Especially, DNA methylation is a major epigenetic modification, and its role is well studied in organ development and cell differentiation [9–11]. Generally, DNA methylation of the promoter regions causes suppression of gene expression [12]. In mammals, three DNA methyltransferases (Dnmt)—Dnmt1, Dnmt3a, and Dnmt3b—coordinately regulate DNA methylation in the genome. Dnmt1 promotes DNA methylation after DNA replication and plays a major role in the maintenance of methylation [13]. Dnmt3a and Dnmt3b are required for the initiation of de novo DNA methylation [9]. Dnmt3a and Dnmt3b mRNA expression are detected in the neonatal and adult livers.

The *Gpam* promoter region contains three SREs [6]. Bisulfite analysis revealed high DNA methylation levels of the *Gpam* promoter containing SREs in the neonatal liver. By contrast, less DNA methylation levels were observed in the adult liver. In the primary culture of neonatal mouse hepatocytes, DNA methylation levels of the *Gpam* promoter are high at the time of seeding but gradually decreased in a time-dependent manner. Reciprocally, GPAT1 mRNA levels are increased in the time course of culture, suggesting an inverse correlation between DNA methylation of the *Gpam* promoter and its mRNA expression in the primary culture of neonatal mouse hepatocytes. On the other hand, DNA

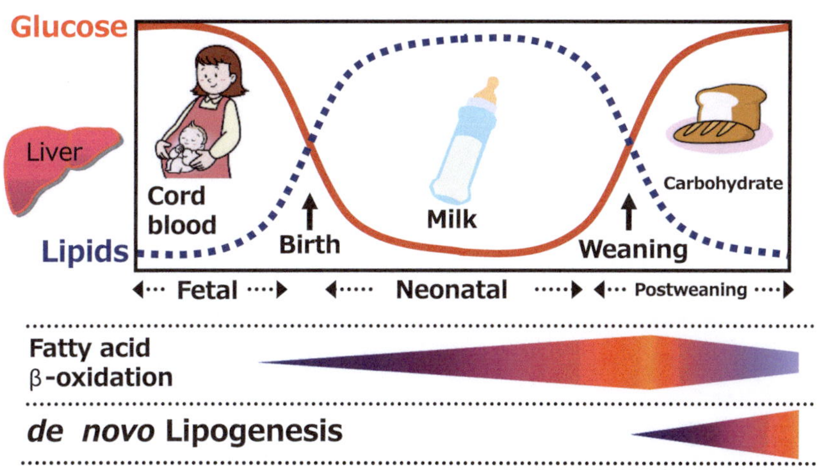

Fig. 1 Changes in hepatic lipid metabolism in response to nutritional environment in mammals. After birth, metabolic pathways such as fatty acid β-oxidation and de novo lipogenesis are activated in the liver as indicated

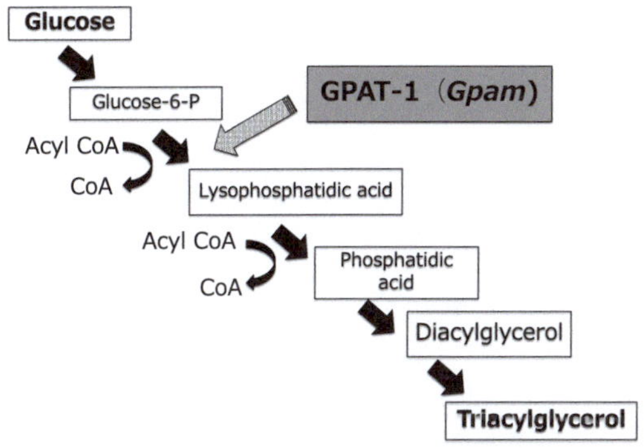

Fig. 2 De novo hepatic lipogenesis and glycerol-3-phosphate acyltransferase 1(GPAT1). Glycerol-3-phosphate acyltransferase 1 (GPAT1) is located on the mitochondrial membrane and a rate-limiting enzyme of de novo hepatic lipogenesis

methylation of the *Scd1* and *Fasn* promoters, containing SREs, is not observed in both the neonatal and adult livers [14]. In contrast, DNA methylation levels are high in the repetitive element of intracisternal A particle (IAP) [14, 15], one of the markers of global DNA methylation, in both the neonatal and adult livers. It is therefore likely that DNA methylation differs in certain regions of the genome between the neonatal and adult livers.

Chromatin immunoprecipitation (ChIP) assays revealed that SREBP-1c is not recruited to SREs in the *Gpam* promoter in the neonatal liver, although it is clearly recruited in the adult liver (Fig. 3). On the other hand, Dnmt3b is strongly recruited to CpG sites (i.e., cytosine followed by guanine) located around SREs in the *Gpam* promoter in the neonatal liver, but not in the adult liver. Thus, it is thought that DNA methylation of the CpG sites in the *Gpam* promoter, which is likely to be induced by Dnmt3b, inhibited the recruitment of SREBP-1c to SREs.

On the other hand after weaning, possibly by nutritional change, DNA methylation of the *Gpam* promoter is decreased allowing the recruitment of SREBP-1c to SREs, and GPAT1 gene expression is activated [14].

2.3 Histone Modification of the *Gpam* Promoter in the Neonatal Liver

Transcriptionally active histone codes such as histone H3 lysine-4-trimethylation (H3K4me3) and lysine-9-acetylation (H3K9Ac) are increased, and the repressive codes such as histone H3 lysine-9-di-methylation (H3K9me2) are decreased at the *Gpam* promoter in the adult liver relative to the neonatal liver (Fig. 3) [14].

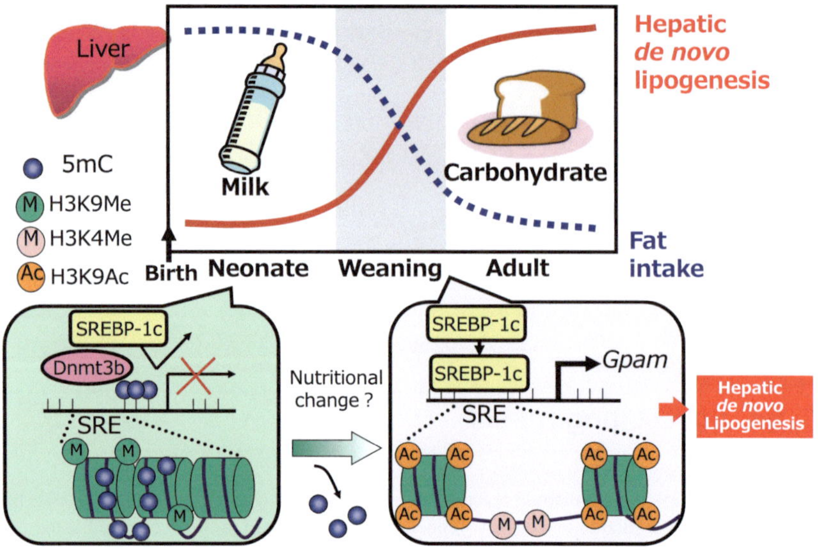

Fig. 3 Epigenetic regulation of GPAT1 gene (*Gpam*). DNA methylation of *Gpam*, which is likely to be induced by DNA methyltransferase (Dnmt)3b, inhibited the recruitment of a major lipogenic transcription factor, sterol regulatory element-binding protein-1c (SREBP-1c). On the other hand, after weaning, possibly by nutritional change, DNA methylation of *Gpam* is decreased allowing the recruitment of SREBP-1c, and *Gpam* expression is activated. 5mC:5-methylated cytosine

2.4 Maternal Nutritional Status Would Affect DNA Methylation Levels of the *Gpam* Promoter

A high-fat or high-calorie diet fed to female animals during gestation and lactation has been reported to increase lipogenic gene expression and TG levels in the liver of the offspring, suggesting that hepatic lipid metabolism is nutritionally affected early in life. In the offspring derived from high-fat and high-calorie diet-fed dams, DNA methylation levels of the *Gpam* promoter are lower, and GPAT1 mRNA levels and SREBP-1 recruitment to SREs are higher than those in the offspring of standard diet-fed dams. Moreover, hepatic TG levels increased in the offspring of high-fat and high-calorie diet-fed dams relative to that of the standard diet-fed dams. These findings suggest that maternal overnutrition could cause decreased DNA methylation of the *Gpam* promoter with increased GPAT1 expression and TG level in the liver of the offspring [14]. However, whether the changes in DNA methylation of the *Gpam* promoter in early life may affect the metabolic phenotypes in later life awaits further investigation.

3 Epigenetic Regulation of Fatty Acid β-Oxidation in the Neonatal Liver

3.1 Genome-Wide DNA Methylation Changes Are Most Induced in the Lactation Period

Analysis of genome-wide DNA methylation changes in the mouse liver revealed that DNA methylation changes are most induced in the lactation period [16]. The genes with an inverse correlation between DNA methylation and mRNA expression, especially those that lost DNA methylation with increased mRNA expression in the lactation period, are significantly related to metabolic function.

3.2 PPARα Is a Key Factor for the DNA Demethylation in the Mouse Liver in the Lactation Period

The binding motif of the nuclear receptor, peroxisome proliferator-activated receptor α (PPARα), which is referred to as PPAR response element (PPRE) characterized by direct repeat 1 (DR1) that consists of two repeats of specific sequence spaced by one nucleotide, is significantly enriched in the genes that lost DNA methylation in the lactation period. The pathway analysis using these gene sets that lost DNA methylation with increased mRNA expression significantly highlighted "PPAR signaling pathway" and "fatty acid metabolism." PPARα is considered as a key regulator of lipid metabolism in the liver, especially hepatic fatty acid β-oxidation [17–19]. Most of the genes related to hepatic fatty acid β-oxidation are PPARα target genes.

Analysis of genome-wide methylation changes in the PPARα-deficient (PPARα-KO) mouse [20] liver revealed that no DNA methylation changes in the perinatal mouse liver. These findings suggest that PPARα also could be a key factor of reduced DNA methylation in the lactation period [16]. However, to date, the molecular mechanism by which PPARα induces reduced DNA methylation remains unclear.

3.3 DNA Demethylation of the Fatty Acid β-Oxidation Genes Is PPARα-Dependent

Representative fatty acid β-oxidation genes, such as acyl-CoA oxidase 1 (*Acox1*) and enoyl-CoA, and hydratase/3-hydroxyacyl CoA dehydrogenase (*Ehhadh*), that are also PPARα target genes show significant decrease in DNA methylation, referred to as DNA demethylation in the lactation period. Moreover, the mRNA expression and protein levels of these genes are significantly increased upon the onset of breast-feeding. There are significant negative correlations between DNA methylation and mRNA expression levels in all the genes. Serum TG concentrations and hepatic TG contents are significantly decreased in the lactation period, which is consistent with the upregulation of the fatty acid β-oxidation in the liver. On the other hand, DNA demethylation of *Acox1* and *Ehhadh* is significantly attenuated in the postnatal PPARα-KO mouse liver. Consistently, mRNA expression levels of *Acox1* and *Ehhadh* of PPARα-KO mice are significantly decreased compared to those of wild-type mice at the time of weaning [16].

3.4 Ligand-Activated PPARα via Breast Milk Accelerates DNA Demethylation of the Fatty Acid β-Oxidation Genes

Maternal administration of Wy14643 (Wy), an artificial ligand of PPARα during the gestation and lactation periods, induces accelerated DNA demethylation of *Acox1* and *Ehhadh* in the liver of the offspring. Accordingly, mRNA expression levels of *Acox1* and *Ehhadh* of the offspring derived from Wy-administered dams (Wy group) are significantly increased compared to those of the offspring derived from vehicle-administered dams (Vehicle group). Serum TG concentrations are significantly decreased, whereas hepatic acetyl-CoA levels are significantly increased, in the Wy group at the time of weaning, suggesting that the fatty acid β-oxidation is activated in the Wy group. These findings indicate that DNA demethylation of the fatty acid β-oxidation genes is ligand-activated PPARα-dependent.

3.5 Significance of DNA Demethylation Via PPARα in the Lactation Period and Clinical Perspective for the Preemptive Medicine

Hepatic fatty acid β-oxidation is activated after birth to produce energy from breast milk lipids [21]. Moreover, The DNA demethylation does not occur in the fetal mouse liver under the physiologic condition, suggesting that it is specific to the lactation period. Therefore, it is thought that

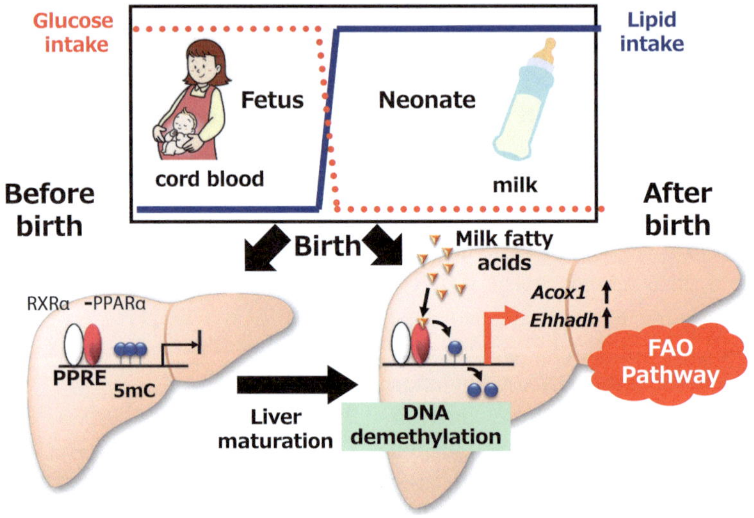

Fig. 4 Ligand-activated PPARα regulates the fatty acid β-oxidation genes through DNA demethylation. Before birth, expression of the fatty acid β-oxidation genes may be suppressed in a DNA methylation-dependent manner, which is partly because PPARα ligands are unavailable under the physiological condition. After birth, upon the onset of breast-feeding, activation of hepatic PPARα by milk lipid ligands may result in the activation of the fatty acid β-oxidation pathway via DNA demethylation. However, the molecular mechanism by which liganded PPARα induces DNA demethylation remains unclear. RXRα, retinoid X receptor α, a nuclear receptor, which is a heterodimeric partner with PPARα on PPRE. *FAO* fatty acid oxidation, *5mC* 5-methylated cytosine

after birth, upon the onset of breast-feeding, activation of hepatic PPARα by milk lipid ligands may result in the activation of the fatty acid β-oxidation pathway via DNA demethylation. PPARα-dependent and lifestage-specific DNA demethylation of the fatty acid β-oxidation genes in the neonatal liver is an adaptive response to the changes in nutritional environment at the onset of lactation (Fig. 4) [16]. These findings also suggest that the onset of lactation, that is a drastic nutritional change, could be an environmental cue to initiate epigenetic modification; DNA demethylation of the fatty acid β-oxidation genes to adapt the nutritional environment thereafter.

Given that the DNA methylation status established in early life remains relatively stable throughout life and thus affects hepatic lipid metabolism and susceptibility to lifestyle diseases such as obesity and type 2 diabetes in later life [22], the development of formula milk with ingredients to activate PPARα and intervention to adequate mother-child nutrition may offer therapeutic strategies in early life to prevent the development of metabolic diseases in adulthood as a preemptive medicine.

References

1. Perez-Castillo A, Schwartz HL, Oppenheimer JH. Rat hepatic mRNA-S14 and lipogenic enzymes during weaning: role of S14 in lipogenesis. Am J Phys. 1987;253:E536–42.
2. Decaux JF, Ferré P, Robin D, Robin P, Girard J. Decreased hepatic fatty acid oxidation at weaning in the rat is not linked to a variation of malonyl-CoA concentration. J Biol Chem. 1988;263:3284–9.
3. Wendel AA, Lewin TM, Coleman RA. Glycerol-3-phosphate acyltransferases: rate limiting enzymes of triacylglycerol biosynthesis. Biochim Biophys Acta. 1791;2009:501–6.
4. Lindén D, William-Olsson L, Ahnmark A, et al. Liver-directed overexpression of mitochondrial glycerol-3-phosphate acyltransferase results in hepatic steatosis, increased triacylglycerol secretion and reduced fatty acid oxidation. FASEB J. 2006;20:434–43.
5. Hammond LE, Gallagher PA, Wang S, et al. Mitochondrial glycerol-3-phosphate acyltransferase-deficient mice have reduced weight and liver triacylglycerol content and altered glycerolipid fatty acid composition. Mol Cell Biol. 2002;22:8204–14.
6. Ericsson J, Jackson SM, Kim JB, Spiegelman BM, Edwards PA. Identification of glycerol-3-phosphate acyltransferase as an adipocyte determination and differentiation factor 1- and sterol regulatory element-binding protein responsive gene. J Biol Chem. 1997;272:7298–305.

7. Tabor DE, Kim JB, Spiegelman BM, Edwards PA. Identification of conserved cis-elements and transcription factors required for sterol-regulated transcription of stearoyl-CoA desaturase 1 and 2. J Biol Chem. 1999;274:20603–10.
8. Wong RH, Chang I, Hudak CS, Hyun S, Kwan HY, Sul HS. A role of DNA-PK for the metabolic gene regulation in response to insulin. Cell. 2009;136:1056–72.
9. Okano M, Bell DW, Haber DA, Li E. DNA methyltransferases Dnmt3a and Dnmt3b are essential for de novo methylation and mammalian development. Cell. 1999;99:247–57.
10. Weber M, Hellmann I, Stadler MB, et al. Distribution, silencing potential and evolutionary impact of promoter DNA methylation in the human genome. Nat Genet. 2007;39:457–66.
11. Mikkelsen TS, Xu Z, Zhang X, et al. Comparative epigenomic analysis of murine and human adipogenesis. Cell. 2010;143:156–69.
12. Maunakea AK, Nagarajan RP, Bilenky M, et al. Conserved role of intragenic DNA methylation in regulating alternative promoters. Nature. 2010;466:253–7.
13. Sharif J, Muto M, Takebayashi S, et al. The SRA protein Np95 mediates epigenetic inheritance by recruiting Dnmtl to methylated DNA. Nature. 2007;450:908–12.
14. Ehara T, Kamei Y, Takahashi M, et al. Role of DNA methylation in the regulation of lipogenic glycerol-3-phosphate acyltransferase 1 gene expression in the mouse neonatal liver. Diabetes. 2012;61:2442–50.
15. Kato Y, Kaneda M, Hata K, et al. Role of the Dnmt3 family in de novo methylation of imprinted and repetitive sequences during male germ cell development in the mouse. Hum Mol Genet. 2007;16:2272–80.
16. Ehara T, Kamei Y, Yuan X, et al. Ligand-activated PPARα-dependent DNA demethylation regulates the fatty acid β-oxidation genes in the postnatal liver. Diabetes. 2015;64:775–84.
17. Lefebvre P, Chinetti G, Fruchart JC, Staels B. Sorting out the roles of PPAR alpha in energy metabolism and vascular homeostasis. J Clin Invest. 2006;116:571–80.
18. Rakhshandehroo M, Knoch B, Muller M, Kersten S. Peroxisome proliferator activated receptor alpha target genes. PPAR Res. 2010;2010
19. Kersten S, Desvergne B, Wahli W. Roles of PPARs in health and disease. Nature. 2000;405:421–4.
20. Lee SS, Pineau T, Drago J, et al. Targeted disruption of the alpha isoform of the peroxisome proliferator-activated receptor gene in mice results in abolishment of the pleiotropic effects of peroxisome proliferators. Mol Cell Biol. 1995;15:3012–22.
21. Périchon R, Bourre JM. Peroxisomal beta-oxidation activity and catalase activity during development and aging in mouse liver. Biochimie. 1995;77:288–93.
22. Gluckman PD, Hanson MA, Buklijas T, Low FM, Beedle AS. Epigenetic mechanisms that underpin metabolic and cardiovascular diseases. Nat Rev Endocrinol. 2009;5:401–8.

Part II

Disease and Environment

Developmental Origins of Nonalcoholic Fatty Liver Disease (NAFLD)

Hiroaki Itoh and Naohiro Kanayama

Abstract

Nonalcoholic fatty liver disease (NAFLD) is a hepatic manifestation of metabolic syndrome. Its prevalence is currently increasing not only in developed obese countries but also in developing countries. Recent findings from human cohorts and animal studies suggest that a nutritional imbalance in the early critical period is causatively associated with the incidence of NAFLD in later life. Based on the current theory of the developmental origins of health and disease (DOHaD), undernourishment and overnourishment in utero are both hypothesized to prime the predisposition for hepatic fat storage. Current knowledge on the developmental origins of NAFLD is introduced in this chapter.

Keywords

Metabolic syndrome · Birth weight · Obesity · Nutrition · Pregnancy · Developmental origins of health and disease (DOHaD) · Fatty liver

H. Itoh (✉) · N. Kanayama
Department of Obstetrics and Gynecology, Hamamatsu University School of Medicine, Higashi-ku, Hamamatsu, Japan
e-mail: hitou-endo@umin.ac.jp

1 Introduction

Nonalcoholic fatty liver disease (NAFLD) is a hepatic manifestation of metabolic syndrome [1] and the most common liver disease not only in adults but also in children, particularly in Western countries [2, 3]. NAFLD indicates a spectrum of pathology from simple hepatic steatosis to more severe nonalcoholic steatohepatitis (NASH). The natural history of NASH in pediatrics currently remains unknown; however, approximately one third of adult patients with NASH are expected to have progressive inflammation and fibrosis that ultimately result in cirrhosis within several to 10 years [4].

The prevalence of NAFLD is currently increasing not only in developed but also in developing countries [5]. Its prevalence in the general population has been estimated to be 20–30% in Western countries and 5–18% in developing countries, such as Asia [6]. However, the rate of increases in the prevalence of NAFLD is markedly greater in developing than in developed countries, and this has been attributed to the widespread availability of obesogenic, cheap, and energy-dense foods [7, 8]. The prevalence of NAFLD in China and Japan has nearly doubled in the last 10–15 years [1, 6, 7, 9, 10]. To the best of our knowledge, it has not yet been established why the prevalence of NAFLD has been simultaneously and rapidly increasing not only in developed but also in developing countries.

© Springer Nature Singapore Pte Ltd. 2018

T. Kubota, H. Fukuoka (eds.), *Developmental Origins of Health and Disease (DOHaD)*, Advances in Experimental Medicine and Biology 1012, https://doi.org/10.1007/978-981-10-5526-3_4

Obesity is strongly associated with the incidence of NAFLD [5, 11–13]. More than two thirds of severely obese people have NAFLD, while more than one half of NAFLD patients are obese in developed countries [5, 11–13]. However, obese individuals do not always have NAFLD, whereas nonobese individuals may develop NAFLD [14, 15]. Kojima et al. previously reported that approximately one half of Japanese patients with NAFLD were nonobese with a body mass index (BMI) of less than 25 kg/m^2 [9]. Thus, the pathophysiological background for the rapidly increasing prevalence of NAFLD in developing and developed countries and the common natural history of NAFLD itself have not yet been elucidated in detail.

The clinical discovery of NAFLD in children has led to speculation that its origins lie in environmental insults or exposure in earlier life [16]. Nobili et al. suggested the programming hypothesis of pro-steatotic conditions presumably caused by undernourishment or overnourishment in utero as a risk factor for NAFLD in later life [17, 18]. This is consistent with the current theory of the developmental origins of health and disease (DOHaD) that environmental factors, including nutritional conditions, during the fetal, neonatal, and infantile periods are related to the risk of noncommunicable diseases (NCDs), such as metabolic syndrome, in later life [19–23]. In this chapter, we discuss the possible contribution of pro-steatotic conditions hypothetically programmed by undernourishment or overnourishment in utero to the increasing prevalence of NAFLD and in developing and obese developed countries, respectively, from the standpoint of the DOHaD theory.

2 Small Newborns and Risk of NAFLD in Later Life

The findings of several studies on humans suggest that small neonates are predisposed to liver disease, including NAFLD, in later life [5, 24–26]. Fraser et al. revealed a relationship between low birth weight and elevated liver enzymes by a random sampling of 2101 British women aged 60–79 years, indicating a possible developmental risk of hepatic cellular injury in these subjects [24]. An epidemiological study on 1587 aged participants by the Helsinki Birth Cohort Study showed that birth and childhood body sizes were negatively associated with NAFLD outcomes after adjustments for adult BMI and also that individuals who had been small in early life and obese as adults were at the highest risk of developing NAFLD [26]. Faienza et al. reported that NAFLD was observed in 34.8% of children who were born as small for gestational age (SGA), but not in those as appropriate for gestational age (AGA) [27]. Alisi et al. detected SGA in 38.9% of children complicated with NAFLD and in 6.7% of uncomplicated children [28]. Moreover, Nobili et al. showed that the prevalence of NASH was higher in SGA children with NAFLD [25]. However, Breij et al. demonstrated that rapid catch-up growth in the first 3 months after birth rather than small birth weight per se was a stronger risk factor for NAFLD in the early adult period [29].

Collectively, these findings suggest that small newborns are predisposed to NAFLD in later life, particularly if they exhibit rapid catch-up growth soon after birth; however, the numbers of studies as well as those of their participants were limited.

3 Undernourishment in Dams and Hepatic Steatosis in Pups

In animal models, general reduction in energy supply to dams is one of the most commonly used methods to establish undernourishment in utero and fetal growth restriction in mammals [5, 30–32]. We and other researchers reported that maternal global nutrient restriction primes the deterioration of hepatic steatosis in the adult offspring of ovine [33], rats [34], and mice [35]. George et al. reported that aged ovine female offspring born to mothers that received global nutrient restriction in the first half of gestation showed significantly increased hepatic lipid and glycogen contents, concomitant with greater weight gain [33]. Magee et al. found that 9-month-old male rat offspring from dams that received 50% global nutrient restriction in the latter half of gestation showed significantly increased hepatic triglyceride contents [34]. The same research group reported that hepatic fat deposition occurred in fetuses exposed to maternal undernutrition, as early as embryonic day 20, prior to the development of offspring adiposity in their rat animal model [36]. We developed a mouse animal model of fetal undernutrition by reducing maternal global nutrient supply in the latter half of the pregnancy period [37], the adult offspring of which showed not only the deterioration of hepatic steatosis [35] but also a wide variety of phenotypical shifts, such as hypothalamic changes resulting in the exacerbation of general obesity [37], the augmentation of macrophage infiltration in white adipose tissue concomitant with increased local fat deposition [38], and the angiotensin II-related deterioration of cardiac remodeling [39, 40]. We recently reported that the severity of maternal energy restriction in this animal model differently primes patterns for the localization of fat deposits in adult offspring [10]. Relatively mild maternal energy restriction, i.e., 30–35%, induced an increase in liver and body weights in adult offspring, while more severe maternal energy restriction, i.e., 40%, elevated liver weights, but not body weights in adult offspring [10, 35]. Since the accumulation of fat in the liver from the entire body was speculated in the latter condition of the animal model, we speculated that the severity of undernourishment in utero may be, at least partly in particular cases, associated with the incidence of NAFLD in non-obese individuals.

In addition to global energy restriction, other animal models have identified factors leading to the exacerbation of hepatic steatosis in later life. Maternal dietary protein restriction during pregnancy and lactation in rat dams was reported to aggravate hepatic steatosis in adult offspring without a parallel increase in general adiposity [41, 42]. Offspring born to rat dams subjected to vitamin B12 and folate deficiencies developed hepatic steatosis at weaning [43]. As a causative factor other than nutritional conditions, Maeyama et al. showed that restraint stress in pregnant mouse dams induced hepatic steatosis in pups at weaning [44].

Thus, studies using animal models have indicated that an insufficient maternal energy supply in pregnancy leads to the trait of preferential fat deposition in the livers of adult offspring. As described in the previous section, findings obtained from humans suggest that small neonates are predisposed to NAFLD. Collectively, information obtained from human and animal studies have led us to the hypothetical concept that undernourishment in utero primes the predisposition for NAFLD in later life. However, caution is needed regarding this scientific logic because small newborns are not always exposed to undernourishment in utero.

4 Maternal Obesity and the Risk of NAFLD in Offspring

In the past few decades, there has been increasing human evidence to suggest that maternal obesity or high maternal glucose levels in pregnancy contribute to offspring obesity and metabolic syndrome [5, 45–47]. NAFLD is regarded as a hepatic manifestation of metabolic syndrome [48]; however, a direct relationship between maternal obesity and the risk of NAFLD in the offspring of humans has been demonstrated in recent years [3, 5]. Brumbaugh et al. reported a 68% increase in intrahepatocellular lipid contents in newborns born to obese mothers with gestational diabetes (GDM) using magnetic resonance imaging (MRI) [49]. The relationship between maternal BMI and neonatal hepatic fat was previously reported to be independent of neonatal subcutaneous fat, suggesting that fetal hepatic fat storage accumulated through excessive maternal free fatty acids via a pathway that is distinct from that of adipose tissue [49]. Modi et al. found a positive correlation between maternal BMI and intrahepatocellular lipid contents in infants measured by MRI [50]. Patel et al. demonstrated that maternal diabetes/glycosuria during pregnancy as well as a higher maternal prepregnancy BMI correlated with the incidence of NAFLD, diagnosed by ultrasound, in offspring at a mean age of 17.8 years [51].

These findings from human studies suggest that maternal obesity and/or hyperglycemia during the critical early periods of developmental plasticity increase susceptibility to and the severity of NAFLD in offspring; however, the number of human studies has been limited.

5 Overnourishment in Dams and Hepatic Steatosis in Pups

In contrast to the limited number of human cohort studies, numerous experimental animal models, such as nonhuman primates, rats, and mice, using a wide variety of dietary approaches have provided detailed and extensive evidence linking the maternal condition of overnutrition to the development of hepatic steatosis in offspring [3, 5, 52]. A chronic maternal high-fat diet (HFD) before and during pregnancy and in the lactation period was found to induce hepatic steatosis in offspring independent of offspring obesity in nonhuman primate [53] and mouse [54] animal models. McCurdy et al. demonstrated that changing the maternal diet to a low-fat diet in subsequent pregnancies improved hepatic steatosis in offspring, indicating that HFD during pregnancy plays a critical role in the programming of offspring hepatic fat deposition [53]. This has been consistently observed in other dietary animal models of various kinds of obesogenic chow diets [55–57].

On the other hand, Thorn et al. chronically supplied HFD to nonhuman primate mothers and found that offspring from insulin-resistant mothers but not insulin-sensitive females developed significant hepatic steatosis even though they consumed a healthy diet after weaning independently of obesity [16]. Thus, maternal hyperglycemia during pregnancy may induce hepatic steatosis in offspring presumably via fetal hyperglycemia in addition to maternal HFD [16].

Kruse et al. reported that offspring exposed to HFD during pregnancy and the lactation period were more susceptible to the development of simple hepatic steatosis, even if they consumed a normal chow diet after weaning, suggesting that pregnancy and lactation periods are the critical windows for HFD-induced susceptibility to hepatic steatosis [58]. Thus, the supply of HFD to dams before and during pregnancy and in the lactation period induced hepatic steatosis in the pups; however, further exposure of their offspring to postweaning HFD was shown to promote the progression of hepatic steatosis to the stage of steatohepatitis [54, 59–61].

6 Developmental Origins of NAFLD from the DOHaD Theory

Numerous epidemiological studies on different ethnic groups in different areas worldwide as well as the identification of some mechanistic backgrounds by excellent animal studies [21, 22, 62] have developed into the novel concept that the establishment of metabolic regulation during the early critical period affects the morbidity of noncommunicable diseases (NCDs) throughout life, i.e., DOHaD [20, 23, 63]. Since epidemiological studies have revealed that undernourishment and overnourishment in utero are both causatively associated with the risk of NCDs in later life, a 'U-shaped curve' theory was proposed for the relationship between nutritional conditions in utero and the risk of developing NCDs in later life [30, 32, 64–67]. As described in the previous sections, evidence from several human cohorts as well as excellent supportive animal studies suggest that undernourishment and overnourishment in utero both prime the risk of NAFLD in later life, suggesting that the 'U-shaped curve' theory is also applicable to the developmental origins of NAFLD (Fig. 1).

Among large numbers of theoretical models for the DOHaD theory concerning undernourishment in utero, the *thrifty phenotype* hypothesis by Hales and Barker [68] is one of the most promising and reasonable sounding models to explain the mechanistic basis underlying possible relationships between undernourishment in utero and obesity-related metabolic disorders in later life [66, 69–71]. They proposed the concept of an adaptive response to undernourishment in utero that is a

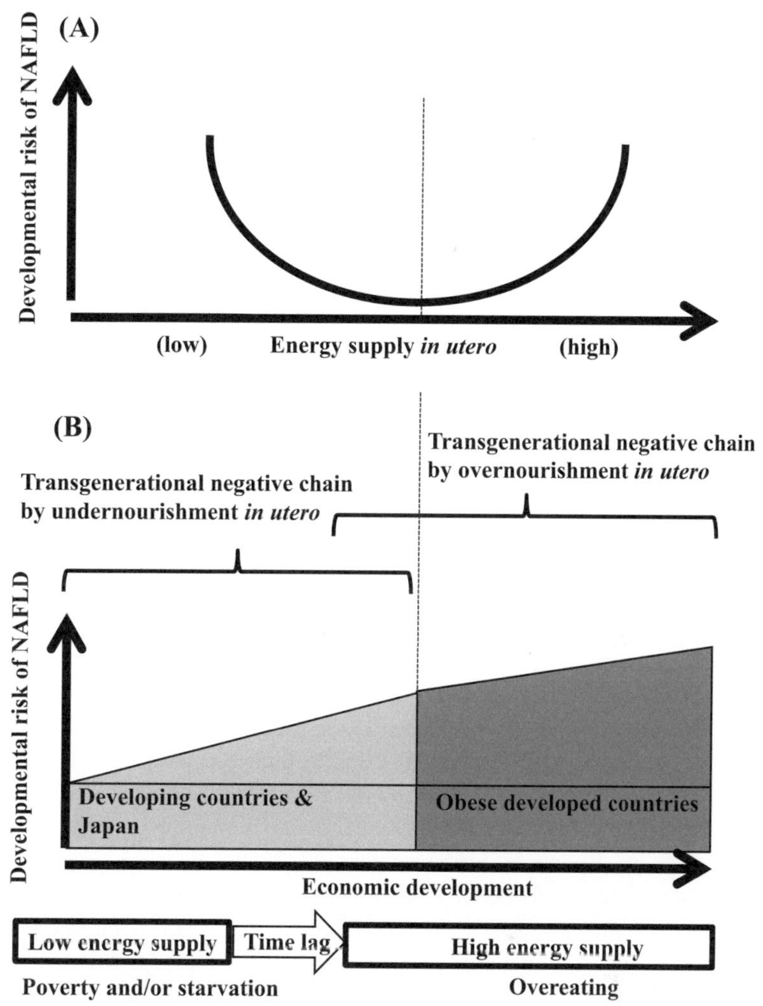

Fig. 1 A two-hit model of nonalcoholic fatty liver disease (NAFLD). 1st hit: energy supply in utero and the developmental risk of NAFLD in later life (**a**). 2nd hit: a possible relationship with economic development stages (**b**)

trade-off between saving energy consumption in utero and downsizing the fetal body. Embryonic and/or fetal *predictive adaptive responses* may adjust the development of their own metabolic regulation systems, including a predisposition to hepatic fat deposition, in response to the environmental characteristics surrounding their mothers for the purpose of *matching* themselves to the prenatally predicted postnatal circumstances of long-lasting insufficient food supply and improving their survivability through a life of incessant starvation [68, 71, 72].

The prevalence of NAFLD is currently increasing in developing countries and obese developed countries [5]. However, the rate of increases in the prevalence of NAFLD was found to be markedly higher in developing countries than in developed countries and has been attributed to the widespread availability of obesogenic, cheap, and energy-dense foods [7, 8]. The prevalence of NAFLD in China and Japan has nearly doubled in the last 10–15 years [1, 6, 7, 9, 10]. Developing countries have been undergoing rapid and prominent economic improvements over the past few decades, and generations subjected to a low energy supply during fetal life due to poverty and/or confusion in society have now shifted to a life with an obesogenic diet [32]. Therefore, individuals expected to have acquired the *thrifty phenotype* in utero encounter a *mismatch* to the excess energy supply provided by a calorie-rich obesogenic diet and develop the risk of NAFLD due to their acquired predisposition to hepatic fat deposition (Fig. 1). Humans have evolved to adapt to continuous starvation for hundreds of thousands of years; however, those in developing countries now need to adjust to an unexpected drastic environmental shift from undernourishment in early life to overnourishment in later life in a single generation. Future economic and political reconstructions of current conflict areas such as the Middle East or Africa may also be a cause for future concern because a cheap obesogenic diet may be consumed by those exposed to a low nutritional environment during fetal life.

Kojima et al. demonstrated that the prevalence of NAFLD was 12.6% in Japan in 1989 and gradually increased to 30.3% in 1998, indicating a 2.4-fold increase over 12 years [9]. Nonobese individuals with a BMI of less than 25 kg/m^2 were previously reported to account for approximately half of all Japanese patients with NAFLD [9]. In Japan, undernourishment is common among women of childbearing age due to their strong desire to be thin [73–76] and has resulted in an increase in low birth weight neonates and a decrease in the mean birth weight [73–76]. Kubota et al. reported that the mean energy intake by pregnant Japanese women was less than 1600 kilocalories/day throughout pregnancy and was 30% (second trimester) and 37% (third trimester) less than the recommendations by the Ministry of Health, Labour and Welfare of Japan, indicating large numbers of relatively undernourished Japanese fetuses due to insufficient maternal energy intake [77]. In consideration of the particular nutrient conditions of young Japanese women, it is plausible that the expected prevalence of relative undernourishment in Japanese fetuses by low maternal energy intake may be, at least partly, related to recent and rapid increases in the prevalence of NAFLD in Japanese individuals, including nonobese patients [10]. Long-term cohort studies including data on individual energy intakes during pregnancy are needed in order to prove this logic.

It appears to be reasonable that the pathophysiological DOHaD hypothesis for undernourishment in utero, such as a *predictive adaptive response*, *thrifty phenotype*, and *mismatch*, is applicable to understand the pathophysiology of the developmental origins of NAFLD associated with undernourishment in utero. Nevertheless, a pathophysiological theory has not yet been established to explain the contribution of overnourishment in utero to the prevalence of NAFLD in later life. It currently remains unclear whether a permanent phenotypic shift in the predisposition to hepatic fat storage in response to overnourishment in utero is advantageous for the survival of the fittest in later life with a permanently excess energy intake. Difficulties are associated with explaining these responses using the *mismatch* hypothesis. Since humans have genetically evolved to adapt to starvation over hundreds of thousands of years, the prevalence of hepatic fat storage in obese developed countries may be a type of *maladaptation* or *functional disorder* induced by a lifelong excess energy supply that they are unable to cope with as a species at present. Nevertheless, it is an urgent task to theorize the relationship between a life-long excess energy supply and the risk of NAFLD, particularly that observed in obese developed countries, in the DOHaD scheme.

Li et al. showed using a rodent model that HFD feeding through three generations progressively induced severe hepatic steatosis in offspring, suggesting transgenerational accumulation in the deterioration of hepatic steatosis by maternal HFD [78]. Moreover, several animal studies suggested that HFD to dams not only primes simple hepatic steatosis in the pups but also increases susceptibility to steatohepatitis, especially when HFD is also supplied to the pups after weaning [54, 59–61]. Since an obesogenic diet is closely linked with individual family eating habits in humans, it is a cause for concern that large numbers of infants from obese families may be subjected to habitual obesogenic meals soon after weaning and, as a result, may be more susceptible to not only NAFLD but also NAS in later life. A human cohort study is needed in order to prove this speculation. However, these animal studies and the wide prevalence of energy-

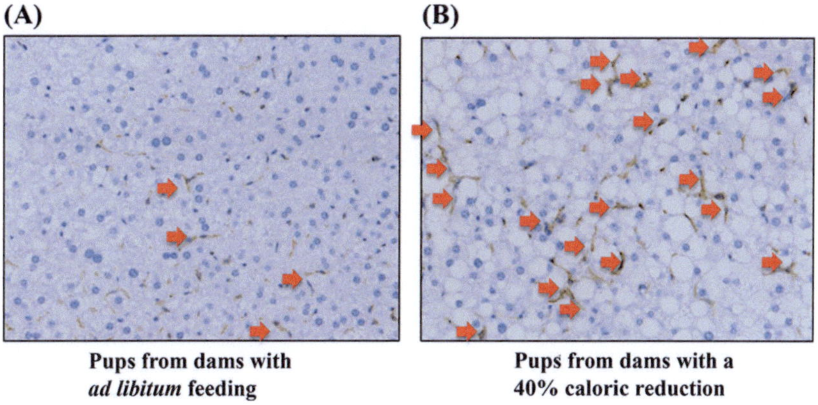

Fig. 2 Infiltration of hepatic macrophages in mouse pups from dams with ad libitum feeding (**a**) and dams with a 40% reduced energy supply (**b**). Immunostaining of macrophage-specific F4/80 in the livers of pups at 17 weeks of age [35]. Yellow staining indicates positive immunostaining. Red arrows indicate hepatic macrophages

dense obesogenic diets together suggest that there is a transgenerational negative chain of NAFLD prevalence in obese developed countries (Fig. 1).

As described in the previous sections, undernourishment and overnourishment in utero both prime the risk of NAFLD in later life (Fig. 1). Previous studies suggested that the developmental risk of hepatic steatosis is independent of that of general adiposity [10, 49, 53, 54], suggesting a specific as well as independent mechanism in the developmental origins of NAFLD. Nevertheless, how these two contradictory fetal environments, i.e., an excess and insufficient energy supply during fetal life, prime the common hepatic phenotype of exaggerated fat deposition, NAFLD, in later life currently remains unknown. We and other researchers showed using rodent animal models that undernourishment in utero induced hepatic chronic inflammation concomitant with an augmentation in hepatic fat storage in later life [34, 35]. Figure 2 shows the enhanced infiltration of hepatic macrophages primed by undernourishment in utero [35]. Overnourishment in utero was reported to induce hepatic chronic inflammation in the offspring of animal models [3, 16, 59, 79]. Since chronic inflammation plays a pivotal role in the development of hepatic steatosis [80–82], we currently speculate the presence of common priming pathways not only for hepatocytes but also hepatic macrophages by undernourishment and overnourishment in utero, constituting the first hit for developing NAFLD (Fig. 3). More intensive studies are needed in order to prove this hypothesis.

7 Conclusion

The prevalence of NAFLD is rapidly increasing in developing and obese developed countries. Recent findings from human cohort and animal studies suggest that the different stages of socioeconomic development are causatively associated with the prevalence of NAFLD via different types of imbalances in nutritional supply in the early critical period of developmental plasticity. It would be a promising research target to highlight the nutritional circumstances in early life as a starting point for the risk accumulation of NAFLD from the standpoint of the DOHaD theory, for the purpose of developing the early intervention strategy of preemptive medicine [83] against the current NAFLD prevalence.

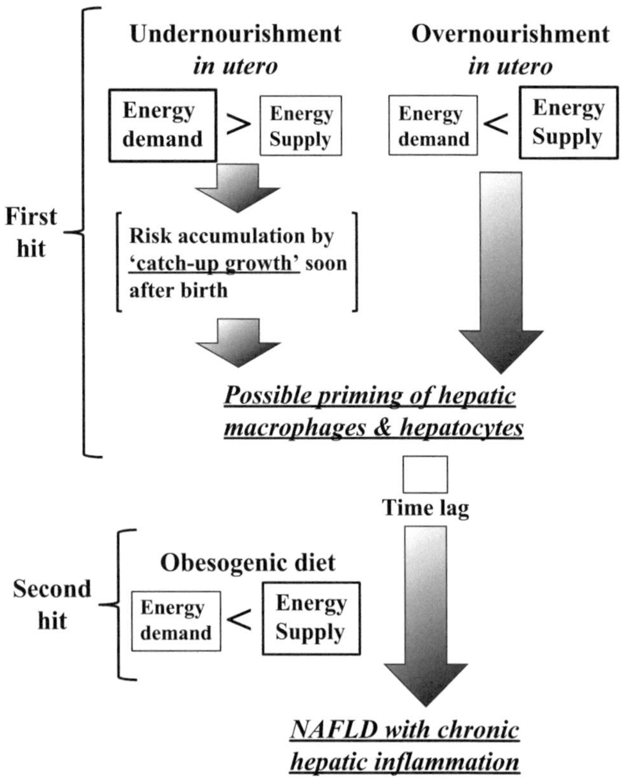

Fig. 3 Hypothetical priming of NAFLD by undernourishment or overnourishment in utero

Acknowledgments This work was supported in part by Grants-in-Aid for Scientific Research from the Ministry of Education, Science, Culture and Sports, Japan (No 15H04882, No.16 K15703).

Disclosures We have no conflict of interest.

References

1. Eguchi Y, Hyogo H, Ono M, Mizuta T, Ono N, Fujimoto K, Chayama K, et al. Prevalence and associated metabolic factors of nonalcoholic fatty liver disease in the general population from 2009 to 2010 in Japan: a multicenter large retrospective study. J Gastroenterol. 2012;47:586–95.
2. Welsh JA, Karpen S, Vos MB. Increasing prevalence of nonalcoholic fatty liver disease among United States adolescents, 1988–1994 to 2007–2010. J Pediatr. 2013;162:496–500 e491.
3. Stewart MS, Heerwagen MJ, Friedman JE. Developmental programming of pediatric nonalcoholic fatty liver disease: redefining the "first hit". Clin Obstet Gynecol. 2013;56:577–90.
4. Argo CK, Northup PG, Al-Osaimi AM, Caldwell SH. Systematic review of risk factors for fibrosis progression in non-alcoholic steatohepatitis. J Hepatol. 2009;51:371–9.
5. Li M, Reynolds CM, Segovia SA, Gray C, Vickers MH. Developmental programming of nonalcoholic fatty liver disease: the effect of early life nutrition on susceptibility and disease severity in later life. Biomed Res Int. 2015;2015:437107.
6. Masarone M, Federico A, Abenavoli L, Loguercio C, Persico M. Non alcoholic fatty liver: epidemiology and natural history. Rev Recent Clin Trials. 2014;9:126–33.
7. Fan JG, Saibara T, Chitturi S, Kim BI, Sung JJ, Chutaputti A. Asia-Pacific working party for N what are the risk factors and settings for non-alcoholic fatty liver disease in Asia-Pacific? J Gastroenterol Hepatol. 2007;22:794–800.
8. Karbasi-Afshar R, Saburi A, Khedmat H. Cardiovascular disorders in the context of non-alcoholic fatty liver disease: a literature review. J Tehran Heart Cent. 2014;9:1–8.
9. Kojima S, Watanabe N, Numata M, Ogawa T, Matsuzaki S. Increase in the prevalence of fatty liver in Japan over the past 12 years: analysis of clinical background. J Gastroenterol. 2003;38:954–61.
10. Itoh H, Muramatsu-Kato K, Ferdous UJ, Kohmura-Kobayashi Y, Kanayama N. Undernourishment in utero and hepatic steatosis in later life; a potential issue in Japanese people: a review. Congenit Anom (Kyoto). 2016;57:178–83.

11. Bellentani S, Saccoccio G, Masutti F, Croce LS, Brandi G, Sasso F, Cristanini G, et al. Prevalence of and risk factors for hepatic steatosis in northern Italy. Ann Intern Med. 2000;132:112–7.
12. Angulo P, Keach JC, Batts KP, Lindor KD. Independent predictors of liver fibrosis in patients with nonalcoholic steatohepatitis. Hepatology. 1999;30:1356–62.
13. Luyckx FH, Desaive C, Thiry A, Dewe W, Scheen AJ, Gielen JE, Lefebvre PJ. Liver abnormalities in severely obese subjects: effect of drastic weight loss after gastroplasty. Int J Obes Relat Metab Disord. 1998;22:222–6.
14. Liu CJ. Prevalence and risk factors for non-alcoholic fatty liver disease in Asian people who are not obese. J Gastroenterol Hepatol. 2012;27:1555–60.
15. Feng RN, Du SS, Wang C, Li YC, Liu LY, Guo FC, Sun CH. Lean-non-alcoholic fatty liver disease increases risk for metabolic disorders in a normal weight Chinese population. World J Gastroenterol. 2014;20:17932–40.
16. Thorn SR, Baquero KC, Newsom SA, El Kasmi KC, Bergman BC, Shulman GI, Grove KL, et al. Early life exposure to maternal insulin resistance has persistent effects on hepatic NAFLD in juvenile nonhuman primates. Diabetes. 2014;63:2702–13.
17. Nobili V, Cianfarani S, Agostoni C. Programming, metabolic syndrome, and NAFLD: the challenge of transforming a vicious cycle into a virtuous cycle. J Hepatol. 2010;52:788–90.
18. Cianfarani S, Agostoni C, Bedogni G, Berni Canani R, Brambilla P, Nobili V, Pietrobelli A. Effect of intrauterine growth retardation on liver and long-term metabolic risk. Int J Obes. 2012;36:1270–7.
19. Gluckman PD, Hanson MA. Developmental origins of health and disease. Cambridge: Cambridge University Press; 2006.
20. Gluckman PD, Hanson MA. Living with the past: evolution, development, and patterns of disease. Science. 2004;305:1733–6.
21. Hanson M, Godfrey KM, Lillycrop KA, Burdge GC, Gluckman PD. Developmental plasticity and developmental origins of non-communicable disease: theoretical considerations and epigenetic mechanisms. Prog Biophys Mol Biol. 2011;106:272–80.
22. Barouki R, Gluckman PD, Grandjean P, Hanson M, Heindel JJ. Developmental origins of non-communicable disease: implications for research and public health. Environ Health. 2012;11:42.
23. Hanson MA, Gluckman PD. Developmental origins of health and disease – Global public health implications. Best Pract Res Clin Obstet Gynaecol. 2014;29:24–31.
24. Fraser A, Ebrahim S, Smith GD, Lawlor DA. The associations between birthweight and adult markers of liver damage and function. Paediatr Perinat Epidemiol. 2008;22:12–21.
25. Nobili V, Marcellini M, Marchesini G, Vanni E, Manco M, Villani A, Bugianesi E. Intrauterine growth retardation, insulin resistance, and nonalcoholic fatty liver disease in children. Diabetes Care. 2007;30:2638–40.
26. Sandboge S, Perala MM, Salonen MK, Blomstedt PA, Osmond C, Kajantie E, Barker DJ, et al. Early growth and non-alcoholic fatty liver disease in adulthood-the NAFLD liver fat score and equation applied on the Helsinki birth cohort study. Ann Med. 2013;45:430–7.
27. Faienza MF, Brunetti G, Ventura A, D'Aniello M, Pepe T, Giordano P, Monteduro M, et al. Nonalcoholic fatty liver disease in prepubertal children born small for gestational age: influence of rapid weight catch-up growth. Horm Res Paediatr. 2013;79:103–9.
28. Alisi A, Panera N, Agostoni C, Nobili V. Intrauterine growth retardation and nonalcoholic fatty liver disease in children. Int J Endocrinol. 2011;2011:269853.
29. Breij LM, Kerkhof GF, Hokken-Koelega AC. Accelerated infant weight gain and risk for non-alcoholic fatty liver disease in early adulthood. J Clin Endocrinol Metab. 2014;99:1189–95.
30. Taylor PD, Poston L. Developmental programming of obesity in mammals. Exp Physiol. 2007;92:287–98.
31. Jones RH, Ozanne SE. Fetal programming of glucose-insulin metabolism. Mol Cell Endocrinol. 2009;297:4–9.
32. Itoh H, Kanayama N. Nutritional conditions in early life and risk of non-communicable diseases (NCDs); the perspective of preemptive medicine in perinatal care. Hypertens Res Preg. 2015;3:1–12.
33. George LA, Zhang L, Tuersunjiang N, Ma Y, Long NM, Uthlaut AB, Smith DT, et al. Early maternal undernutrition programs increased feed intake, altered glucose metabolism and insulin secretion, and liver function in aged female offspring. Am J Physiol Regul Integr Comp Physiol. 2012;302:R795–804.
34. Magee TR, Han G, Cherian B, Khorram O, Ross MG, Desai M. Down-regulation of transcription factor peroxisome proliferator-activated receptor in programmed hepatic lipid dysregulation and inflammation in intrauterine growth-restricted offspring. Am J Obstet Gynecol. 2008;199:271 e271–75.
35. Muramatsu-Kato K, Itoh H, Kohmura-Kobayashi Y, Ferdous UJ, Tamura N, Yaguchi C, Uchida T, et al. Undernourishment in utero primes hepatic steatosis in adult mice offspring on an obesogenic diet; involvement of endoplasmic reticulum stress. Sci Rep. 2015;5:16867.
36. Yamada M, Wolfe D, Han G, French SW, Ross MG, Desai M. Early onset of fatty liver in growth-restricted rat fetuses and newborns. Congenit Anom (Kyoto). 2011;51:167–73.
37. Yura S, Itoh H, Sagawa N, Yamamoto H, Masuzaki H, Nakao K, Kawamura M, et al. Role of premature leptin surge in obesity resulting from intrauterine undernutrition. Cell Metab. 2005;1:371–8.
38. Kohmura YK, Kanayama N, Muramatsu K, Tamura N, Yaguchi C, Uchida T, Suzuki K, et al. Association between body weight at weaning and remodeling in the subcutaneous adipose tissue of obese adult mice with undernourishment in utero. Reprod Sci. 2013;20:813–27.
39. Kawamura M, Itoh H, Yura S, Mogami H, Suga S, Makino H, Miyamoto Y, et al. Undernutrition in utero

augments systolic blood pressure and cardiac remodeling in adult mouse offspring: possible involvement of local cardiac angiotensin system in developmental origins of cardiovascular disease. Endocrinology. 2007;148:1218–25.
40. Kawamura M, Itoh H, Yura S, Mogami H, Fujii T, Kanayama N, Konishi I. Angiotensin II receptor blocker candesartan cilexetil, but not hydralazine hydrochloride, protects against mouse cardiac enlargement resulting from undernutrition in utero. Reprod Sci. 2009;16:1005–12.
41. Souza-Mello V, Mandarim-de-Lacerda CA, Aguila MB. Hepatic structural alteration in adult programmed offspring (severe maternal protein restriction) is aggravated by post-weaning high-fat diet. Br J Nutr. 2007;98:1159–69.
42. Erhuma A, Salter AM, Sculley DV, Langley-Evans SC, Bennett AJ. Prenatal exposure to a low-protein diet programs disordered regulation of lipid metabolism in the aging rat. Am J Physiol Endocrinol Metab. 2007;292:E1702–14.
43. Pooya S, Blaise S, Moreno Garcia M, Giudicelli J, Alberto JM, Gueant-Rodriguez RM, Jeannesson E, et al. Methyl donor deficiency impairs fatty acid oxidation through PGC-1alpha hypomethylation and decreased ER-alpha, ERR-alpha, and HNF-4alpha in the rat liver. J Hepatol. 2012;57:344–51.
44. Maeyama H, Hirasawa T, Tahara Y, Obata C, Kasai H, Moriishi K, Mochizuki K, et al. Maternal restraint stress during pregnancy in mice induces 11beta-HSD1-associated metabolic changes in the livers of the offspring. J Dev Orig Health Dis. 2015;6:105–14.
45. Whitaker RC. Predicting preschooler obesity at birth: the role of maternal obesity in early pregnancy. Pediatrics. 2004;114:e29–36.
46. Boney CM, Verma A, Tucker R, Vohr BR. Metabolic syndrome in childhood: association with birth weight, maternal obesity, and gestational diabetes mellitus. Pediatrics. 2005;115:e290–6.
47. Shankar K, Harrell A, Liu X, Gilchrist JM, Ronis MJ, Badger TM. Maternal obesity at conception programs obesity in the offspring. Am J Physiol Regul Integr Comp Physiol. 2008;294:R528–38.
48. Fotbolcu H, Zorlu E. Nonalcoholic fatty liver disease as a multi-systemic disease. World J Gastroenterol. 2016;22:4079–90.
49. Brumbaugh DE, Tearse P, Cree-Green M, Fenton LZ, Brown M, Scherzinger A, Reynolds R, et al. Intrahepatic fat is increased in the neonatal offspring of obese women with gestational diabetes. J Pediatr. 2013;162:930–936 e931.
50. Modi N, Murgasova D, Ruager-Martin R, Thomas EL, Hyde MJ, Gale C, Santhakumaran S, et al. The influence of maternal body mass index on infant adiposity and hepatic lipid content. Pediatr Res. 2011;70:287–91.
51. Patel S, Lawlor DA, Callaway M, Macdonald-Wallis C, Sattar N, Fraser A. Association of maternal diabetes/glycosuria and pre-pregnancy body mass index with offspring indicators of non-alcoholic fatty liver disease. BMC Pediatr. 2016;16:47.
52. Vos MB. Furthering the understanding of maternal obesity in nonalcoholic fatty liver disease. Hepatology. 2013;58:4–5.
53. McCurdy CE, Bishop JM, Williams SM, Grayson BE, Smith MS, Friedman JE, Grove KL. Maternal high-fat diet triggers lipotoxicity in the fetal livers of nonhuman primates. J Clin Invest. 2009;119:323–35.
54. Bruce KD, Cagampang FR, Argenton M, Zhang J, Ethirajan PL, Burdge GC, Bateman AC, et al. Maternal high-fat feeding primes steatohepatitis in adult mice offspring, involving mitochondrial dysfunction and altered lipogenesis gene expression. Hepatology. 2009;50:1796–808.
55. Bayol SA, Simbi BH, Fowkes RC, Stickland NC. A maternal "junk food" diet in pregnancy and lactation promotes nonalcoholic fatty liver disease in rat offspring. Endocrinology. 2010;151:1451–61.
56. Pruis MG, Lendvai A, Bloks VW, Zwier MV, Baller JF, de Bruin A, Groen AK, et al. Maternal western diet primes non-alcoholic fatty liver disease in adult mouse offspring. Acta Physiol (Oxf). 2014;210:215–27.
57. Kjaergaard M, Nilsson C, Rosendal A, Nielsen MO, Raun K. Maternal chocolate and sucrose soft drink intake induces hepatic steatosis in rat offspring associated with altered lipid gene expression profile. Acta Physiol (Oxf). 2014;210:142–53.
58. Kruse M, Seki Y, Vuguin PM, Du XQ, Fiallo A, Glenn AS, Singer S, et al. High-fat intake during pregnancy and lactation exacerbates high-fat diet-induced complications in male offspring in mice. Endocrinology. 2013;154:3565–76.
59. Mouralidarane A, Soeda J, Visconti-Pugmire C, Samuelsson AM, Pombo J, Maragkoudaki X, Butt A, et al. Maternal obesity programs offspring nonalcoholic fatty liver disease by innate immune dysfunction in mice. Hepatology. 2013;58:128–38.
60. Gregorio BM, Souza-Mello V, Carvalho JJ, Mandarim-de-Lacerda CA, Aguila MB. Maternal high-fat intake predisposes nonalcoholic fatty liver disease in C57BL/6 offspring. Am J Obstet Gynecol. 2010;203:495 e491–98.
61. Oben JA, Mouralidarane A, Samuelsson AM, Matthews PJ, Morgan ML, McKee C, Soeda J, et al. Maternal obesity during pregnancy and lactation programs the development of offspring nonalcoholic fatty liver disease in mice. J Hepatol. 2010;52:913–20.
62. Haugen AC, Schug TT, Collman G, Heindel JJ. Evolution of DOHaD: the impact of environmental health sciences. J Dev Orig Health Dis. 2015;6:55–64.
63. Silveira PP, Portella AK, Goldani MZ, Barbieri MA. Developmental origins of health and disease (DOHaD). J Pediatr. 2007;83:494–504.
64. Curhan GC, Willett WC, Rimm EB, Spiegelman D, Ascherio AL, Stampfer MJ. Birth weight and adult hypertension, diabetes mellitus, and obesity in US men. Circulation. 1996;94:3246–50.

65. Fernandez-Twinn DS, Ozanne SE. Mechanisms by which poor early growth programs type-2 diabetes, obesity and the metabolic syndrome. Physiol Behav. 2006;88:234–43.
66. Prentice AM, Moore SE. Early programming of adult diseases in resource poor countries. Arch Dis Child. 2005;90:429–32.
67. Katz AR. Noncommunicable diseases: global health priority or market opportunity? An illustration of the World Health Organization at its worst and at its best. Int J Health Serv. 2013;43:437–58.
68. Hales CN, Barker DJ. The thrifty phenotype hypothesis. Br Med Bull. 2001;60:5–20.
69. Muhlhausler BS, Adam CL, McMillen IC. Maternal nutrition and the programming of obesity: the brain. Organogenesis. 2008;4:144–52.
70. Prentice AM, Rayco-Solon P, Moore SE. Insights from the developing world: thrifty genotypes and thrifty phenotypes. Proc Nutr Soc. 2005;64:153–61.
71. Wells JC. The thrifty phenotype: an adaptation in growth or metabolism? Am J Hum Biol. 2011;23:65–75.
72. Gluckman PD, Hanson MA. The fetal matrix -evolution, development and disease. Cambridge: Cambridge University Press; 2005. p. 78–102.
73. Goto Y. Diseases in the 21st century. J Jpn Soc Study of Obes (Jpn). 2006;12:1–2.
74. Hayashi F, Takimoto H, Yoshita K, Yoshiike N. Perceived body size and desire for thinness of young Japanese women: a population-based survey. Br J Nutr. 2006;96:1154–62.
75. Gluckman PD, Seng CY, Fukuoka H, Beedle AS, Hanson MA. Low birthweight and subsequent obesity in Japan. Lancet. 2007;369:1081–2.
76. Itoh H, Kanayama N. Low birthweight and risk of obesity -potential problem of Japanese people. Curr Women Health Rev. 2009;5:212–9.
77. Kubota K, Itoh H, Tasaka M, Naito H, Fukuoka Y, Muramatsu Kato K, Kohmura YK, et al. Changes of maternal dietary intake, bodyweight and fetal growth throughout pregnancy in pregnant Japanese women. J Obstet Gynaecol Res. 2013;39:1383–90.
78. Li J, Huang J, Li JS, Chen H, Huang K, Zheng L. Accumulation of endoplasmic reticulum stress and lipogenesis in the liver through generational effects of high fat diets. J Hepatol. 2012;56:900–7.
79. Ashino NG, Saito KN, Souza FD, Nakutz FS, Roman EA, Velloso LA, Torsoni AS, et al. Maternal high-fat feeding through pregnancy and lactation predisposes mouse offspring to molecular insulin resistance and fatty liver. J Nutr Biochem. 2012;23:341–8.
80. Hotamisligil GS. Inflammation and metabolic disorders. Nature. 2006;444:860–7.
81. Hotamisligil GS. Inflammation and endoplasmic reticulum stress in obesity and diabetes. Int J Obes. 2008;32(Suppl 7):S52–4.
82. Than NN, Newsome PN. A concise review of non-alcoholic fatty liver disease. Atherosclerosis. 2015;239:192–202.
83. Imura H. Life course health care and preemptive approach to non-communicable diseases. Proc Jpn Acad Ser B Phys Biol Sci. 2013;89:462–73.

Fetal Origins of Hypertension

Yuichiro Arima, Koichi Nishiyama,
Yasuhiro Izumiya, Koichi Kaikita, Seiji Hokimoto,
and Kenichi Tsujita

Abstract

Hypertension is a common noncommunicable disease. According to the World Health Organization, 1.13 billion people were suffering from hypertension in the year 2015. High blood pressure, hypertension, has a multifactorial etiology. Arterial atherosclerotic changes, systolic or diastolic dysfunction of the heart, and other noncardiac factors are involved. Epidemiological evidence has revealed that perinatal growth disturbance elevates the prevalence of hypertension. However, the specific effects of developmental disturbances on the pathological process of hypertension are poorly understood. Recently, it has become apparent that the perinatal period plays many essential roles in cardiovascular development. In this chapter, we focus on the perinatal development of the cardiovascular system, especially in murine models. Individual organs, blood, blood vessels, and the heart show unique growth characteristics during this period. We also introduce evidence from related clinical studies regarding the developmental origins of hypertension. Finally, evidence from several animal models is presented to reveal the effects of developmental disturbance or stress on arterial pathology. Improving our understanding of both developmental events and the results of clinical studies will give fresh insight into the fetal origins of hypertension.

Keywords

Hypertension · Cardiovascular Development

1 Introduction

Hypertension is a common noncommunicable disease that is marked by constantly elevated arterial blood pressure. The presence of hypertension also increases the risk of other noncommunicable diseases such as ischemic heart disease, arrhythmia, cerebrovascular diseases, and so on. Blood pressure can be expressed as the product of cardiac output and systemic vascular resistance, similar to Ohm's law (E (blood pressure) = I (cardiac output) × R (systemic vascular resistance)) (Fig. 1). Cardiac output and systemic vascular resistance are influenced by many

Y. Arima (✉)
Department of Cardiovascular Medicine, Kumamoto University, Kumamoto, Japan

International Research Center for Medical Science, Kumamoto University, Kumamoto, Japan
e-mail: arimay@kumamoto-u.ac.jp

K. Nishiyama
International Research Center for Medical Science, Kumamoto University, Kumamoto, Japan

Y. Izumiya · K. Kaikita · S. Hokimoto · K. Tsujita
Department of Cardiovascular Medicine, Kumamoto University, Kumamoto, Japan

© Springer Nature Singapore Pte Ltd. 2018

T. Kubota, H. Fukuoka (eds.), *Developmental Origins of Health and Disease (DOHaD)*, Advances in Experimental Medicine and Biology 1012, https://doi.org/10.1007/978-981-10-5526-3_5

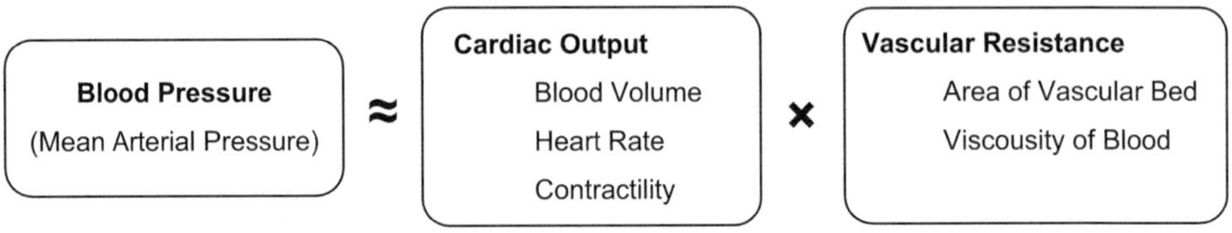

Fig. 1 Major determinants of blood pressure. Blood pressure can be represented as the product of cardiac output and vascular resistance. Each cardiac output or vascular resistance is also influenced by many factors.

Table 1 Developmental events during perinatal period

Organ	Events
Blood	Sift of the hematopoietic fields (AGM-liver-BM) Expansion and differentiation of hematopoietic stem cells
Vessels	Organ-specific vascular development (e.g., retina) Maturation of remodeled vessels
Heart	Sift of the cardiomyocyte metabolisms Cardiomyocyte maturation Coronary vasculature development

AGM Aorta-gonado-mesonephros, *BM* bone marrow

factors; therefore, a variety of alterations of the cardiovascular system are involved in the progression of hypertension.

In the past, the perinatal period was regarded simply as a phase of enlargement of existing organs. However, recent evidence has revealed that perinatal development of the cardiovascular system plays a unique and pivotal role in organ growth and maturation (Table 1).

In this chapter, we first focus on the murine perinatal development of each component of the cardiovascular system: the blood, blood vessels, and the heart. Recent clinical findings on the developmental origins of hypertension are then introduced. We also discuss several animal models that capture disturbances of perinatal development.

2 The Early Stage of Cardiovascular Development

During development, the heart and blood vessels are derived from the same origin, i.e., the lateral plate mesoderm. The lateral plate mesoderm is found at the periphery of the notochord and divided into two parts: the somatic mesoderm and the splanchnic mesoderm. The splanchnic mesoderm is the common origin of the heart, the blood vessels, and the blood. The anterior portion of the splanchnic mesoderm is specified to the cardiogenic mesoderm, whereas the posterior portion differentiates into the hemogenic mesoderm.

Organogenesis of the cardiovascular system is complex and beyond the scope of this chapter. The gross architecture of the cardiovascular system, including the four chambers of the heart and the primitive circulatory system, is almost completely formed during the first trimester.

3 Perinatal Cardiovascular Development

3.1 Perinatal Development of the Blood

The hematopoietic field shifts to various organs during the maturation process of hematopoietic stem cells [1]. The yolk sac is known to be the primary hemogenic field. This field produces immature hematopoietic stem cells that migrate into other hemogenic fields, including the aorta-gonado-mesonephros (AGM) region and the placenta. The AGM region is placed at the frontal side of the developing aorta and produces stem cells, which play an important role in specification of hematopoietic stem cells. Hematopoietic stem cells then translocate to the fetal liver at approximately embryonic day 11, where expansion and some differentiation of hematopoietic stem cells occur. The bone marrow starts to develop when the bone tissue has just formed, at approximately embryonic day 14, and continues until the neonatal period [2, 3]. Fetal hematopoietic stem cells change their phenotype to adult stem cells. These changes mainly proceed during the late developmental stage and the postnatal period [4].

A particular subset of endothelial cells is known to acquire hematopoietic potential and to produce hematopoietic stem cells. This population is called the hemogenic endothelium [5]. Hemogenic endothelial cells are produced at limited periods and sites, but their production is conserved among vertebrates. Disturbances of their production cause a lethal effect during embryogenesis [6].

3.2 Perinatal Development of the Vessels

Blood vessel formation is broadly divided into two phenomena, vasculogenesis and angiogenesis [7, 8]. Vasculogenesis is the de novo formation of the primary vascular plexus, and angiogenesis is the emergence of new blood vessels from the preexisting plexus. The timing of vascular formation differs for each organ. The great vessels are already formed by the mid-gestational stage, whereas the retinal vascular network starts to develop in the postnatal period [9].

Mature blood vessels have three layers, the tunica intima, the tunica media, and the tunica adventitia. The tunica intima is the innermost layer and is composed of monolayered endothelial cells. Smooth muscle cells covered by the endothelial layer form the tunica media. The outermost layer, the tunica adventitia, is composed of many cell types. It includes nerves, capillaries (called the vasa vasorum), and connective tissue. The tunica adventitia also contains stem cells, which can differentiate into smooth muscle cells and endothelial cells [10].

3.3 Perinatal Development of the Heart

3.3.1 Caridomyocyte Metabolism

The heart is an energy-consuming organ. To produce the huge amounts of ATP required, the heart can utilize various substrates such as carbohydrates, lipids amino acids, and ketone bodies [11]. In adults, oxidative phosphorylation is the major pathway for cardiac metabolism, and free fatty acids are used as the main substrate.

However, during the fetal stage, glycolysis is the main pathway for energy production. The perinatal stage is a metabolic transition phase, from glycolysis to mitochondrial oxidation [12, 13].

3.3.2 Regenerative Capacity

Previously, after birth, the heart was regarded as a nondividing organ. However, recent reports have shown that regenerative capacity persists in adults of lower vertebrates and during the neonatal period of mammals. This period is closely correlated with the maturation of cardiomyocytes [14–16].

In murine models, proliferation of cardiomyocytes reaches its peak at approximately embryonic day 10 and gradually decreases [17] but persists during the neonatal period [18–20]. Recently, Naqvi et al. reported that a proliferative burst occurs in preadolescent mice [21]. However, another group refuted their findings [22], and the exact timing of the termination of cardiomyocyte proliferation remains debatable.

Binucleation or polyploidy of cardiomyocytes occurs when maturation proceeds [23]. Many factors influence the proliferation and maturation of cardiomyocytes [24, 25]. Neuregulin1 promotes cardiomyocyte proliferation and regeneration through the ErbB4 receptor [26]. Recently, Puente et al. reported that the oxygen concentration induces cardiomyocyte cell cycle termination [27].

3.3.3 Development of the Coronary Vasculature

The coronary vasculature is important for continuous beating of the heart. Stenosis or occlusion of a coronary artery in an adult induces severe ischemic heart diseases, such as angina pectoris and myocardial infarction. The coronary vasculature is composed of cells from many sources, and its development occurs during the perinatal period [28, 29]. A particular cluster of coronary endothelial cells was reported to dramatically increase during the postnatal period [30].

4 Clinical Studies Related to the Developmental Origins of Hypertension

David Barker reported that there is a relationship between birth weight and cardiovascular disease [31]. Subsequently, it was demonstrated that the perinatal environment affects the development of many noncommunicable diseases [32]. Many studies have also reported the involvement of hypertension. The Nurses' Health Study (NHS), in which more than 150,000 women were enrolled, showed that a low birth weight increased the risk of hypertension [33]. A study of the Northern Finland Birth Cohort 1966, in which birth weight and blood pressure at the age of 31 years were recorded, showed that birth weight was inversely associated with blood pressure [34]. A study of the Northern Finland Birth Cohort 1986 also showed the same tendency in systolic blood pressure at 16 years old, especially in girls [35]. The Young Finns Study demonstrated that preterm, or small for gestational age, individuals showed a higher prevalence of adult hypertension [36]. The Bogalusa Heart Study also showed that a low birth weight was associated with systolic blood pressure variability [37].

In addition to condition at birth, postnatal development also affects noncommunicable diseases. Perng et al. reported that rapid gain in body mass index during the first postnatal 6 months increased blood pressure in mid-childhood [38]. Furthermore, faster growth in either weight or height is associated with higher blood pressure in childhood [35].

Several historical events also affirm this concept. The Dutch famine is one of the most well-known life-threatening famines caused by World War II. A follow-up study revealed that survivors showed increased risks of noncommunicable diseases [39]. The Biafran famine, another severe famine, was prompted by the Nigerian Civil War (1967–1970). A cohort study showed that perinatal starvation was associated with elevated risks of hypertension and other noncommunicable diseases [40]. Recently, a large genome-wide study showed that birth weight was inversely correlated with systolic and diastolic blood pressure [41]. Based on these reports, the perinatal environment strongly affects development of hypertension and other noncommunicable diseases.

The hyperfiltration theory addresses the well-known etiology of hypertension in individuals with perinatal growth disturbances. Brenner argued that a lower number of glomeruli cause filtration overload and lead to glomerular damage. Glomerular damage reduces the number of glomeruli, perpetuating the cycle [42]. Intrauterine growth retardation or preterm birth results in a reduced nephron number, and this histological change causes glomerular hyperfiltration [43]. The mechanism will be described in detail in a later chapter.

Various animal models have been used to understand the mechanisms underlying these phenomena. In sheep, embolization of the feeding artery to the placenta or a partial uterine restriction causes placental insufficiency on fetal development. This procedure produces fetuses that are lighter than controls; maturation and cell cycle activities of cardiomyocytes were also suppressed [44, 45]. In rodents, partial calorie or protein restriction during pregnancy is also used to generate intrauterine growth retardation models. Proliferation of cardiomyocytes was repressed in a rat model [46]. It was also reported in humans that infants who were small for their gestational age showed lower left ventricular output and diastolic dysfunction [47]. It is known that the number of cardiomyocytes relates to the contractility of the heart. The postnatal compensation mechanisms of cardiomyocytes might predispose these individuals to hypertension.

Environmental stress causes many types of epigenetic modifications. Cardiac methylome analysis revealed dynamic changes in DNA methylation status among embryonic, neonatal, healthy adult, and failing hearts [48, 49]. In their reports, Gilsbach et al. revealed prototypical cardiomyocyte genes, *myf6* and *myf7*, which were demethylated in newborn cardiomyocytes. Furthermore, neonatal methylation patterns partially resembled to the pressure overloaded-failing heart [48]. These models indicate that the environmental stress can change the methylation status of cardiomyocytes in adult.

Here, we have provided an overview of the perinatal development of the cardiovascular system. We have also introduced the available clinical evidence and provided data from some studies of animal models that are underway. It is almost unknown how epigenetic modifications during the perinatal period affect the adult heart and diseases. Further direct evidence of the perinatal developmental origins of hypertension is required, along with experimental evidence that illuminates the underlining mechanisms.

References

1. Mikkola HK, Orkin SH. The journey of developing hematopoietic stem cells. Development. 2006;133(19):3733–44. https://doi.org/10.1242/dev.02568.
2. Coskun S, Chao H, Vasavada H, Heydari K, Gonzales N, Zhou X, de Crombrugghe B, Hirschi KK. Development of the fetal bone marrow niche and regulation of HSC quiescence and homing ability by emerging osteolineage cells. Cell Reports. 2014;9(2):581–90. https://doi.org/10.1016/j.celrep.2014.09.013.
3. Coskun S, Hirschi KK. Establishment and regulation of the HSC niche: roles of osteoblastic and vascular compartments. Birth Defects Research Part C, Embryo Today : Reviews. 2010;90(4):229–42. https://doi.org/10.1002/bdrc.20194.
4. Kim I, Saunders TL, Morrison SJ. Sox17 dependence distinguishes the transcriptional regulation of fetal from adult hematopoietic stem cells. Cell. 2007;130(3):470–83. https://doi.org/10.1016/j.cell.2007.06.011.
5. Gritz E, Hirschi KK. Specification and function of hemogenic endothelium during embryogenesis. Cellular and Molecular Life Sciences : CMLS. 2016;73(8):1547–67. https://doi.org/10.1007/s00018-016-2134-0.
6. Antas VI, Al-Drees MA, Prudence AJ, Sugiyama D, Fraser ST. Hemogenic endothelium: a vessel for blood production. The International Journal of Biochemistry & Cell Biology. 2013;45(3):692–5. https://doi.org/10.1016/j.biocel.2012.12.013.
7. Adams RH, Alitalo K. Molecular regulation of angiogenesis and lymphangiogenesis. Nature Reviews Molecular Cell Biology. 2007;8(6):464–78. https://doi.org/10.1038/nrm2183.
8. Carmeliet P. Angiogenesis in life, disease and medicine. Nature. 2005;438(7070):932–6. https://doi.org/10.1038/nature04478.
9. Sapieha P. Eyeing central neurons in vascular growth and reparative angiogenesis. Blood. 2012;120(11):2182–94. https://doi.org/10.1182/blood-2012-04-396846.
10. Kramann R, Goettsch C, Wongboonsin J, Iwata H, Schneider RK, Kuppe C, Kaesler N, Chang-Panesso M, Machado FG, Gratwohl S, Madhurima K, Hutcheson JD, Jain S, Aikawa E, Humphreys BD. Adventitial MSC-like Cells Are Progenitors of Vascular Smooth Muscle Cells and Drive Vascular Calcification in Chronic Kidney Disease. Cell Stem Cell. 2016;19(5):628–42. https://doi.org/10.1016/j.stem.2016.08.001.
11. Kolwicz SC Jr, Purohit S, Tian R. Cardiac metabolism and its interactions with contraction, growth, and survival of cardiomyocytes. Circulation Research. 2013;113(5):603–16. https://doi.org/10.1161/CIRCRESAHA.113.302095.
12. Ellen Kreipke R, Wang Y, Miklas JW, Mathieu J, Ruohola-Baker H. Metabolic remodeling in early development and cardiomyocyte maturation. Seminars in cell & Developmental Biology. 2016;52:84–92. https://doi.org/10.1016/j.semcdb.2016.02.004.
13. Lopaschuk GD, Jaswal JS. Energy metabolic phenotype of the cardiomyocyte during development, differentiation, and postnatal maturation. Journal of Cardiovascular Pharmacology. 2010;56(2):130–40. https://doi.org/10.1097/FJC.0b013e3181e74a14.

14. Jopling C, Sleep E, Raya M, Marti M, Raya A, Izpisua Belmonte JC. Zebrafish heart regeneration occurs by cardiomyocyte dedifferentiation and proliferation. Nature. 2010;464(7288):606–9. https://doi.org/10.1038/nature08899.
15. Kikuchi K, Holdway JE, Werdich AA, Anderson RM, Fang Y, Egnaczyk GF, Evans T, Macrae CA, Stainier DY, Poss KD. Primary contribution to zebrafish heart regeneration by gata4(+) cardiomyocytes. Nature. 2010;464(7288):601–5. https://doi.org/10.1038/nature08804.
16. Porrello ER, Mahmoud AI, Simpson E, Hill JA, Richardson JA, Olson EN, Sadek HA. Transient regenerative potential of the neonatal mouse heart. Science. 2011;331(6020):1078–80. https://doi.org/10.1126/science.1200708.
17. Toyoda M, Shirato H, Nakajima K, Kojima M, Takahashi M, Kubota M, Suzuki-Migishima R, Motegi Y, Yokoyama M, Takeuchi T. Jumonji downregulates cardiac cell proliferation by repressing cyclin D1 expression. Developmental Cell. 2003;5(1):85–97.
18. Bergmann O, Zdunek S, Felker A, Salehpour M, Alkass K, Bernard S, Sjostrom SL, Szewczykowska M, Jackowska T, Dos Remedios C, Malm T, Andra M, Jashari R, Nyengaard JR, Possnert G, Jovinge S, Druid H, Frisen J. Dynamics of cell generation and turnover in the human heart. Cell. 2015;161(7):1566–75. https://doi.org/10.1016/j.cell.2015.05.026.
19. Darehzereshki A, Rubin N, Gamba L, Kim J, Fraser J, Huang Y, Billings J, Mohammadzadeh R, Wood J, Warburton D, Kaartinen V, Lien CL. Differential regenerative capacity of neonatal mouse hearts after cryoinjury. Developmental Biology. 2015;399(1):91–9. https://doi.org/10.1016/j.ydbio.2014.12.018.
20. Leone M, Magadum A, Engel FB. Cardiomyocyte proliferation in cardiac development and regeneration: a guide to methodologies and interpretations. American Journal of Physiology Heart and Circulatory Physiology. 2015;309(8):H1237–50. https://doi.org/10.1152/ajpheart.00559.2015.
21. Naqvi N, Li M, Calvert JW, Tejada T, Lambert JP, Wu J, Kesteven SH, Holman SR, Matsuda T, Lovelock JD, Howard WW, Iismaa SE, Chan AY, Crawford BH, Wagner MB, Martin DI, Lefer DJ, Graham RM, Husain A. A proliferative burst during preadolescence establishes the final cardiomyocyte number. Cell. 2014;157(4):795–807. https://doi.org/10.1016/j.cell.2014.03.035.
22. Alkass K, Panula J, Westman M, Wu TD, Guerquin-Kern JL, Bergmann O. No evidence for cardiomyocyte number expansion in preadolescent mice. Cell. 2015;163(4):1026–36. https://doi.org/10.1016/j.cell.2015.10.035.
23. Paradis AN, Gay MS, Zhang L. Binucleation of cardiomyocytes: the transition from a proliferative to a terminally differentiated state. Drug Discovery Today. 2014;19(5):602–9. https://doi.org/10.1016/j.drudis.2013.10.019.
24. Foglia MJ, Poss KD. Building and re-building the heart by cardiomyocyte proliferation. Development. 2016;143(5):729–40. https://doi.org/10.1242/dev.132910.
25. Uygur A, Lee RT. Mechanisms of cardiac regeneration. Developmental Cell. 2016;36(4):362–74. https://doi.org/10.1016/j.devcel.2016.01.018.
26. Bersell K, Arab S, Haring B, Kuhn B. Neuregulin1/ErbB4 signaling induces cardiomyocyte proliferation and repair of heart injury. Cell. 2009;138(2):257–70. https://doi.org/10.1016/j.cell.2009.04.060.
27. Puente BN, Kimura W, Muralidhar SA, Moon J, Amatruda JF, Phelps KL, Grinsfelder D, Rothermel BA, Chen R, Garcia JA, Santos CX, Thet S, Mori E, Kinter MT, Rindler PM, Zacchigna S, Mukherjee S, Chen DJ, Mahmoud AI, Giacca M, Rabinovitch PS, Aroumougame A, Shah AM, Szweda LI, Sadek HA. The oxygen-rich postnatal environment induces cardiomyocyte cell-cycle arrest through DNA damage response. Cell. 2014;157(3):565–79. https://doi.org/10.1016/j.cell.2014.03.032.
28. Arima Y, Miyagawa-Tomita S, Maeda K, Asai R, Seya D, Minoux M, Rijli FM, Nishiyama K, Kim KS, Uchijima Y, Ogawa H, Kurihara Y, Kurihara H. Preotic neural crest cells contribute to coronary artery smooth muscle involving endothelin signalling. Nature Communications. 2012;3:1267. https://doi.org/10.1038/ncomms2258.
29. Wu B, Zhang Z, Lui W, Chen X, Wang Y, Chamberlain AA, Moreno-Rodriguez RA, Markwald RR, O'Rourke BP, Sharp DJ, Zheng D, Lenz J, Baldwin HS, Chang CP, Zhou B. Endocardial cells form the coronary arteries by angiogenesis through myocardial-endocardial VEGF signaling. Cell. 2012;151(5):1083–96. https://doi.org/10.1016/j.cell.2012.10.023.
30. Tian X, Hu T, Zhang H, He L, Huang X, Liu Q, Yu W, He L, Yang Z, Yan Y, Yang X, Zhong TP, Pu WT, Zhou B. Vessel formation. De novo formation of a distinct coronary vascular population in neonatal heart. Science. 2014;345(6192):90–4. https://doi.org/10.1126/science.1251487.
31. Osmond C, Barker DJ, Winter PD, Fall CH, Simmonds SJ. Early growth and death from cardiovascular disease in women. Bmj. 1993;307(6918):1519–24.
32. Hanson M. The birth and future health of DOHaD. Journal of Developmental Origins of Health and Disease. 2015;6(5):434–7. https://doi.org/10.1017/S2040174415001129.
33. Curhan GC, Chertow GM, Willett WC, Spiegelman D, Colditz GA, Manson JE, Speizer FE, Stampfer MJ. Birth weight and adult hypertension and obesity in women. Circulation. 1996;94(6):1310–5.
34. Jarvelin MR, Sovio U, King V, Lauren L, Xu B, McCarthy MI, Hartikainen AL, Laitinen J, Zitting P, Rantakallio P, Elliott P. Early life factors and blood pressure at age 31 years in the 1966 northern Finland birth cohort. Hypertension. 2004;44(6):838–46. https://doi.org/10.1161/01.HYP.0000148304.33869.ee.

35. Sipola-Leppanen M, Vaarasmaki M, Tikanmaki M, Hovi P, Miettola S, Ruokonen A, Pouta A, Jarvelin MR, Kajantie E. Cardiovascular risk factors in adolescents born preterm. Pediatrics. 2014;134(4):e1072–81. https://doi.org/10.1542/peds.2013-4186.
36. Juonala M, Cheung MM, Sabin MA, Burgner D, Skilton MR, Kahonen M, Hutri-Kahonen N, Lehtimaki T, Jula A, Laitinen T, Jokinen E, Taittonen L, Tossavainen P, Viikari JS, Magnussen CG, Raitakari OT. Effect of birth weight on life-course blood pressure levels among children born premature: the Cardiovascular Risk in Young Finns Study. Journal of Hypertension. 2015;33(8):1542–8. https://doi.org/10.1097/HJH.0000000000000612.
37. Chen W, Srinivasan SR, Yao L, Li S, Dasmahapatra P, Fernandez C, Xu J, Berenson GS. Low birth weight is associated with higher blood pressure variability from childhood to young adulthood: the Bogalusa Heart Study. American Journal of Epidemiology. 2012;176(Suppl 7):S99–105. https://doi.org/10.1093/aje/kws298.
38. Perng W, Rifas-Shiman SL, Kramer MS, Haugaard LK, Oken E, Gillman MW, Belfort MB. Early weight gain, linear growth, and mid-childhood blood pressure: a prospective study in project viva. Hypertension. 2016;67(2):301–8. https://doi.org/10.1161/HYPERTENSIONAHA.115.06635.
39. Roseboom T, de Rooij S, Painter R. The Dutch famine and its long-term consequences for adult health. Early Human Development. 2006;82(8):485–91. https://doi.org/10.1016/j.earlhumdev.2006.07.001.
40. Hult M, Tornhammar P, Ueda P, Chima C, Bonamy AK, Ozumba B, Norman M. Hypertension, diabetes and overweight: looming legacies of the Biafran famine. PloS one. 2010;5(10):e13582. https://doi.org/10.1371/journal.pone.0013582.
41. Horikoshi M, Beaumont RN, Day FR, Warrington NM, Kooijman MN, Fernandez-Tajes J, Feenstra B, van Zuydam NR, Gaulton KJ, Grarup N, Bradfield JP, Strachan DP, Li-Gao R, Ahluwalia TS, Kreiner E, Rueedi R, Lyytikainen LP, Cousminer DL, Wu Y, Thiering E, Wang CA, Have CT, Hottenga JJ, Vilor-Tejedor N, Joshi PK, Boh ET, Ntalla I, Pitkanen N, Mahajan A, van Leeuwen EM, Joro R, Lagou V, Nodzenski M, Diver LA, Zondervan KT, Bustamante M, Marques-Vidal P, Mercader JM, Bennett AJ, Rahmioglu N, Nyholt DR, Ma RC, Tam CH, Tam WH, Group CCHW, Ganesh SK, van Rooij FJ, Jones SE, Loh PR, Ruth KS, Tuke MA, Tyrrell J, Wood AR, Yaghootkar H, Scholtens DM, Paternoster L, Prokopenko I, Kovacs P, Atalay M, Willems SM, Panoutsopoulou K, Wang X, Carstensen L, Geller F, Schraut KE, Murcia M, van Beijsterveldt CE, Willemsen G, Appel EV, Fonvig CE, Trier C, Tiesler CM, Standl M, Kutalik Z, Bonas-Guarch S, Hougaard DM, Sanchez F, Torrents D, Waage J, Hollegaard MV, de Haan HG, Rosendaal FR, Medina-Gomez C, Ring SM, Hemani G, McMahon G, Robertson NR, Groves CJ, Langenberg C, Luan J, Scott RA, Zhao JH, Mentch FD, MacKenzie SM, Reynolds RM, Early Growth Genetics C, Lowe WL, Tonjes A, Stumvoll M, Lindi V, Lakka TA, van Duijn CM, Kiess W, Korner A, Sorensen TI, Niinikoski H, Pahkala K, Raitakari OT, Zeggini E, Dedoussis GV, Teo YY, Saw SM, Melbye M, Campbell H, Wilson JF, Vrijheid M, de Geus EJ, Boomsma DI, Kadarmideen HN, Holm JC, Hansen T, Sebert S, Hattersley AT, Beilin LJ, Newnham JP, Pennell CE, Heinrich J, Adair LS, Borja JB, Mohlke KL, Eriksson JG, Widen E, Kahonen M, Viikari JS, Lehtimaki T, Vollenweider P, Bonnelykke K, Bisgaard H, Mook-Kanamori DO, Hofman A, Rivadeneira F, Uitterlinden AG, Pisinger C, Pedersen O, Power C, Hypponen E, Wareham NJ, Hakonarson H, Davies E, Walker BR, Jaddoe VW, Jarvelin MR, Grant SF, Vaag AA, Lawlor DA, Frayling TM, Smith GD, Morris AP, Ong KK, Felix JF, Timpson NJ, Perry JR, Evans DM, McCarthy MI, Freathy RM. Genome-wide associations for birth weight and correlations with adult disease. Nature. 2016;538(7624):248–52. https://doi.org/10.1038/nature19806.
42. Brenner BM, Lawler EV, Mackenzie HS. The hyperfiltration theory: a paradigm shift in nephrology. Kidney International. 1996;49(6):1774–7.
43. Zohdi V, Sutherland MR, Lim K, Gubhaju L, Zimanyi MA, Black MJ. Low birth weight due to intrauterine growth restriction and/or preterm birth: effects on Nephron Number and Long-Term Renal Health. International Journal of Nephrology. 2012;2012:136942. https://doi.org/10.1155/2012/136942.
44. Louey S, Jonker SS, Giraud GD, Thornburg KL. Placental insufficiency decreases cell cycle activity and terminal maturation in fetal sheep cardiomyocytes. The Journal of Physiology. 2007;580(Pt. 2):639–48. https://doi.org/10.1113/jphysiol.2006.122200.
45. Morrison JL, Botting KJ, Dyer JL, Williams SJ, Thornburg KL, McMillen IC. Restriction of placental function alters heart development in the sheep fetus. American Journal of Physiology Regulatory, Integrative and Comparative Physiology. 2007;293(1):R306–13. https://doi.org/10.1152/ajpregu.00798.2006.
46. Corstius HB, Zimanyi MA, Maka N, Herath T, Thomas W, van der Laarse A, Wreford NG, Black MJ. Effect of intrauterine growth restriction on the number of cardiomyocytes in rat hearts. Pediatric Research. 2005;57(6):796–800. https://doi.org/10.1203/01.PDR.0000157726.65492.CD.
47. Sehgal A, Doctor T, Menahem S. Cardiac function and arterial biophysical properties in small for gestational age infants: postnatal manifestations of fetal programming. The Journal of Pediatrics. 2013;163(5):1296–300. https://doi.org/10.1016/j.jpeds.2013.06.030.

48. Gilsbach R, Preissl S, Gruning BA, Schnick T, Burger L, Benes V, Wurch A, Bonisch U, Gunther S, Backofen R, Fleischmann BK, Schubeler D, Hein L. Dynamic DNA methylation orchestrates cardiomyocyte development, maturation and disease. Nature Communications. 2014;5:5288. https://doi.org/10.1038/ncomms6288.

49. Sim CB, Ziemann M, Kaspi A, Harikrishnan KN, Ooi J, Khurana I, Chang L, Hudson JE, El-Osta A, Porrello ER. Dynamic changes in the cardiac methylome during postnatal development. FASEB Journal : Official Publication of the Federation of American Societies for Experimental Biology. 2015;29(4):1329–43. https://doi.org/10.1096/fj.14-264093.

Involvement of Noncoding RNAs in Stress-Related Neuropsychiatric Diseases Caused by DOHaD Theory

ncRNAs and DOHaD-Induced Neuropsychiatric Diseases

Takahiro Nemoto and Yoshihiko Kakinuma

Abstract

According to the DOHaD theory, low birth weight is a risk factor for various noncommunicable chronic diseases that develop later in life. Noncoding RNAs (ncRNAs), including miRNAs, siRNAs, piRNAs, and lncRNAs, are functional RNA molecules that are transcribed from DNA but that are not translated into proteins. In general, miRNAs, siRNAs, and piRNAs function to regulate gene expression at the transcriptional and posttranscriptional levels. Studying ncRNAs has provided opportunities for new diagnosis and therapeutic knowledge in the endocrinological and metabolic fields as well as cancer biology. In this review, we focus on the roles of miRNAs and lncRNAs in the pathophysiology of stress-related neuropsychiatric diseases, which show abnormal blood hormone levels due to loss of feedback control and/or decreased sensitivity. Numerous recent studies have begun to unveil the importance of ncRNAs in regulation of stress-related hormone levels and functions. We summarize the involvement of abnormal ncRNA expression in the development of stress-related neuropsychiatric diseases based on the DOHaD theory.

Keywords

Fetal malnutrition · Low birth weight · HPA axis · Stress · ncRNA · miRNA · lncRNA

Abbreviations

ACTH	Adrenocorticotropin
ADHD	Attention deficit hyperactivity disorder
CeA	Central nucleus of the amygdala
CRF	Corticotropin-releasing factor
CRF-R	CRF receptor
GR	Glucocorticoid receptor
GRE	Glucocorticoid-responsive element
HFD	High-fat diet
HPA	Hypothalamic-pituitary-adrenal
lncRNA	Long noncoding RNA
miRNA	microRNA
ncRNA	Noncoding RNA
POMC	Proopiomelanocortin
PTSD	Post-traumatic stress disorder
PVN	Paraventricular nucleus of the hypothalamus
UTR	Untranslated region

T. Nemoto (✉) · Y. Kakinuma
Department of Physiology, Nippon Medical School, Tokyo, Japan
e-mail: taknemo@nms.ac.jp

© Springer Nature Singapore Pte Ltd. 2018

T. Kubota, H. Fukuoka (eds.), *Developmental Origins of Health and Disease (DOHaD)*, Advances in Experimental Medicine and Biology 1012, https://doi.org/10.1007/978-981-10-5526-3_6

1 Introduction

Developmental origins of health and disease (DOHaD) is the concept in which environmental factors such as maternal malnutrition, stress, and a deviant lifestyle during development could potentially affect a person's health and increase the risks of various diseases later in life [1–4]. Several epidemiological studies have revealed that malnutrition and stress during fetal development cause low birth weight, which is a risk factor for noncommunicable chronic diseases, such as cardiovascular, metabolic, and neuropsychological diseases in the future. Many low birth weight children catch up as they grow, but some of them grow slowly, and long-term growth retardation may result [5, 6]. About half of the growth-retarded children are often inferior to normally developing children in terms of psychological development, and growth-retarded children often have social and emotional problems even in adulthood [7, 8]. However, verifying the theory of DOHaD using humans is ethically and experimentally difficult. Therefore, in this review, we will introduce some animal and cell biological studies that focus on stress-related diseases that involve noncoding RNA (ncRNAs (microRNAs (miRNAs) and long noncoding RNAs (lncRNAs))), including our data.

2 Stress and Hormone

Stress is a biological and psychological response that occurs when a host encounters a threat. Sensing stress, the hypothalamus (at the base of the brain), which is responsible for a stress response, is activated. As a result, the hypothalamus sends signals to two organs: the pituitary and the adrenal medulla [9]. The stress response by the sympathetic nerve is an immediate response. The transmission of hormones secreted from the hypothalamus (H) to the adrenal gland (A) via the pituitary (P) is called the "HPA axis," which plays an important role in the mechanism of a series of stress reactions. This HPA axis is composed of corticotropin-releasing factor (CRF) in the hypothalamus, adrenocorticotropic hormone (ACTH) in the pituitary gland, and cortisol in the adrenal cortex and is regulated by diurnal rhythm and feedback [10, 11]. Glucocorticoid is a steroid hormone synthesized from cholesterol. The main glucocorticoid in humans is cortisol, but mice and rats produce corticosterone as the main glucocorticoid because the synthesis route of steroid hormones is different in rodents compared to humans. Adrenaline from the adrenal medulla and cortisol from the adrenal cortex raise blood pressure and blood sugar and induce a stress response described as "fight or flight." When the crisis cannot be avoided, "freezing (fright)" may occur in which the whole body becomes stiff. Glucocorticoid production by the adrenal cortex maintains many behavioral reactions including fear-related behaviors [12]. Glucocorticoids are essential for normal fright formation, and blood levels are increased in response to many conditions including fear, novelty, and uncertainty. One neural mechanism in which glucocorticoids regulate phobia-related behaviors is the regulation of CRF gene expression in the central nucleus of the amygdala (CeA), which is critically involved in behavioral responses related to fear [13]. CRF is associated with many phobia-related behaviors and mediates its effect on fear-related behaviors by acting on the CeA. In addition, rats systemically treated with corticosterone show enhanced conditional phobia-induced freezing in association with increased CRF gene expression in the CeA [14]. Thus, the HPA axis is involved not only in fight or flight but also in fright. Furthermore, it is also involved in many internal activities including immunity, feeding, sleep, emotion, reproductive behavior, energy metabolism, etc. Based on these facts, exposure to stress may cause stress-related diseases that are accompanied by various symptoms.

3 Stress and Its Related Diseases

Stress-related diseases include cardiovascular diseases, gastrointestinal diseases, reproductive disorders, infection, metabolic diseases such as obesity and diabetes, neurological diseases such as Alzheimer's disease, mental diseases such as insomnia and depression, and myalgic encephalomyelitis [15–17].

Excess or prolonged exposure to cortisol has been implicated in the development of stress-related diseases including depression, schizophrenia, autistic spectrum disorder (autism), attention deficit hyperactivity disorder (ADHD), and insomnia. Stress activates the HPA axis. Glucocorticoid hormones are consistently increased in severely depressed patients and in rodents subjected to stress [18–20]. Stressful life events and the consequent HPA axis hyperactivity are thought to be a cause of depression.

Glucocorticoid receptors are widely distributed in the brain, including the regions responsible for negative regulation of the HPA axis, such as the hypothalamus and hippocampus. Restraint decreases glucocorticoid receptor mRNA expression in the hypothalamus and the hippocampus [21]. A series of animal studies has shown that feeding a high-fat diet (HFD) increases HPA axis activity; plasma corticosterone concentrations are significantly increased in HFD-fed rats, and HFD feeding also elevates both basal and restraint-induced HPA activity in rats. We previously reported impaired downregulation of stress-induced glucocorticoid receptors in HFD-fed rats. Based on these findings, the HPA system is affected by diet and nutrition, indicating that diet modulates stress responses. These alterations increase the risk of developing neuropsychiatric diseases, and when a person with a risk of diet-induced HPA axis upregulation is exposed to strong stress, development of stress-related neuropsychiatric diseases is possible. However, further examination is necessary to confirm this hypothesis.

Decreases in glucocorticoid receptor expression in the central nervous system are associated with elevated HPA axis activity. Moreover, the hippocampus is an important site of negative feedback of the HPA axis. Prolonged stress is associated with a decrease in corticosterone binding to glucocorticoid receptors in the hippocampus in rats [22]. Thus, dysregulated glucocorticoid receptor downregulation in the hypothalamus and hippocampus may prolong restraint-induced elevation of plasma corticosterone levels.

4 ncRNAs

4.1 microRNAs

miRNAs are single-stranded RNAs of 18–22 nucleotides and are important regulatory molecules in many biological processes. Reduction in the amount of a specific mRNA is caused by binding of the miRNA primarily to the 3′-untranslated region [23]. The process of general miRNA production and action is as follows. Primary miRNA (pri-miRNA) is transcribed from the genome by RNA polymerase II in the nucleus and processed to precursor miRNA by Drosha or DGCR8 [24, 25]. The pre-miRNAs are transported to the cytoplasm by the Ric-GTP-dependent nuclear transport receptor exportin 5, and Dicer and its interaction partner, TAR RNA-binding protein, covert pre-miRNA into double-stranded mature small RNA (miRNA/miRNA* duplexes) of about ~22 nucleotides in length [26, 27]. One strand of this miRNA duplex is assembled into a miRNA-induced silencing complex (miRISC) with Dicer, TAR RNA-binding protein (TRBP), and Argonaute proteins [28]. Mature miRNAs induce complementary mRNA targeting of the miRNA-induced silencing complex, thus regulating gene expression by causing mRNA degradation and repression of translation initiation [29]. Many results, particularly from cancer research, showed that miRNA expression profiling is an important tool for disease diagnosis and treatment [30–33].

Cells can release vesicles containing miRNAs directly from the cell membrane or take up vesicles by endocytosis. Many in vitro studies have reported that such released vesicles can be transferred into acceptor cells. Extracellular vesicles are called microvesicles or exosomes. From these results, quantifying the content of miRNA in blood or urine exosomes has become useful for diagnosis without needed to extract (intracellular) miRNA from tissue [34–36].

Each miRNA has hundreds of targets, producing complexity in the regulation of expression by an individual miRNA. miRNAs are generally thought to function by fine-tuning the expression of multiple targets [37, 38]. An altered expression level of a target gene by one miRNA may be small, but additive synergistic changes may occur when multiple miRNAs regulated expression. However, this idea lacks direct evidence, because such changes cannot be observed in vivo. The underlying mechanisms by which miRNA expression levels are altered remain to be elucidated. Altered methylation of regulatory region of genes is one mechanisms of regulation of gene expression, including regulation by miRNAs.

4.2 Long Noncoding RNA (lncRNAs)

As with mRNAs, lncRNAs are transcribed by polymerase II (most are 5′-capped, polyadenylated, and spliced) but contains fewer exons compared to mRNA and shows a tissue-specific expression pattern [39]. Protein-encoding genes cover about 2% of the human genome. In contrast, the ncRNA genes occupy 62–75% of the genome [40]. ncRNAs are often classified by length. lncRNAs can be as long as over 200 nt or more. lncRNAs differ from mRNAs as follows: the expression level is low, and lncRNAs often remain in the nucleus after transcription; lncRNAs have high cell-type specificity compared with mRNAs; although the sequences of lncRNAs themselves are not conserved evolutionarily, the promoters are conserved and show that synteny, which is physical co-localization of genetic loci on the same chromosome within an individual or species, has the same function [41]. In many cases, lncRNAs are often related to gene expression control via chromatin because of their sequence information and high structural flexibility. lncRNAs bind to DNA by base pairing and can bind to proteins via special secondary structure [42]. Therefore, various protein complexes are recruited to a specific part of DNA [43]. In addition, some lncRNAs act as decoys [44] and inhibits protein-DNA binding [45]. Although several physiological roles for lncRNAs are known [46–49], research in this field has just begun, and future research is desired.

5 miRNAs and lncRNAs in Stress-Related Neuropsychiatric Diseases

Recently, many reports have appeared to show involvement of miRNAs in neuropsychiatric diseases, and measurement of miRNAs is useful as a diagnosis or prognosis tool. Regarding the stress response, Rinaldi et al. reported that exposing mice to acute stress significantly increase let-7a, miR-9, and miR-26a/b expression in the prefrontal cortex [50], which is a key target of the maladaptive response to stress.

Mannironi et al. reported that restraint induces an increase in expression of mineralocorticoid receptors in the amygdala, which is a key region for stress response and a target of glucocorticoids, and is negatively correlated with miR-135a and miR-124 expression [51]. Specifically, restraint decreases miR-135a and miR-124 expression in the amygdala of mice. Overexpression of miR-135a or miR-124 significantly suppresses mineralocorticoid receptor promoter activity and mineralocorticoid receptor expression in vitro.

Regarding stress-related neuropsychiatric diseases, Schmidt et al. systematically reviewed the results of epidemiology and animal experiments on miRNA expression profiles and roles in post-traumatic stress disorder (PTSD) and Alzheimer's disease [52]. In addition, Babenko et al. reviewed the involvement of epigenetic mechanisms, such as miRNA expression, DNA methylation, and histone modifications, on many neuropsychiatric diseases, such as schizophrenia, ADHD, autism, and anxiety- or depression-related disorders, in response to stressful experiences and hostile environmental factors [53].

Issler et al. demonstrated that miR-135, the targets of which are the serotonin transporter and serotonin receptor-1a, is upregulated in the raphe of mice treated with imipramine, a serotonin reuptake inhibitor [54]. Overexpression of miR-135a in serotoninergic neurons reduces anxiety- and depression-like behaviors, and knockdown of miR-135a in the raphe increases anxiety-like behavior. They also demonstrated that miR-135a in the blood and brain is downregulated in depressed patients. They suggested that increasing the levels of miR-135 represses a serotonin transporter and serotonin receptors, causing an increase in serotonin in the synaptic cleft, which is associated with a reduction in depressive symptoms. Their findings may lead to a better understanding of the psychopathology that involves dysregulation of the serotonin system and more effective treatment and/or development of biomarkers.

miR-16 also regulates serotonin transporter gene expression. Chronic treatment with a selective serotonin reuptake inhibitor increases miR-16 levels in the raphe. Baudry et al. proposed that miR-16 contributes to the therapeutic aspect of antidepressants in monoaminergic neurons [55]. Song et al. demonstrated that miR-16 in cerebrospinal fluid but not the blood of patients with major depressive disorder is significantly lower than that in controls and that the level of miR-16 is negatively correlated with Hamilton scores and with the level of miR-16 in the cerebrospinal fluid [56]. Their results suggest that miR-16 participated in the physiopathology of depression by the modulating the serotonin transmitter system in the brain.

Like miRNAs, lncRNAs are also expected to be promising candidate molecules for evaluation of neuropsychiatric diseases. lncRNAs regulate genes in psychiatric diseases such as major depressive disorder, schizophrenia, and autism spectrum disorder. Cui et al. investigated 30,586 human lncRNAs in peripheral monocytes and reported that 6 lncRNAs are significantly downregulated in depressed patients [57]. Wang et al. examined the genome-wide expression levels of lncRNAs and found changes in a total of 3929 lncRNAs, including 2407 that were upregulated and 1522 that were downregulated in the peripheral leukocytes of patients with autism spectrum disorder [58]. Ziats et al. profiled over 33,000 annotated lncRNAs and 30,000 mRNA transcripts from postmortem brain tissue of autistic patients. They reported over 200 differentially expressed lncRNAs, which were enriched for genomic regions containing genes related to neurodevelopment and psychiatric diseases [59]. Thus, the majority of studies have analyzed the expression profile of lncRNAs. However, few studies have shown that lncRNAs act as regulators of the target genes in patients with neurological diseases.

Thus, miRNAs and lncRNAs are involved in the onset and pathology of neuropsychiatric diseases, and measurement of their levels is useful for diagnosis and prognosis.

6 Involvement of miRNAs and lncRNAs on HPA Axis Feedback During the Stress Response

Based on the site where the miRNA is integrated in the regulation of gene expression and function, miRNAs can act as new functional factors in the homeostatic response. miRNAs are thought to be involved in stress signal transduction, modulation, stability, buffering, and feedback signaling. When miRNAs are involved in a negative feedback loop, they suppress the expression of hormones or their receptor genes. In this way, miRNAs are also involved in the stress response. Several studies have reported miRNA expression changes in the brain upon exposure to stress. Volk et al. reported that exposing mice to chronic stress leads to a specific increase in miR-15a levels in the amygdala [60]. Mice expressing reduced levels of miR-15a in the amygdala following exposure to chronic stress exhibit increased anxiety-like behaviors. In humans, exposure to childhood trauma is associated with increased miR-15a levels in peripheral blood. Their data suggest that miR-15a plays an important role in stress adaptation and the pathogenesis of stress-related psychopathologies. Higuchi et al. reported that mice exposed to chronic stress exhibit increased depression-like behaviors and reduced hippocampal expression of miR-124 [61]. They also showed that histone deacetylase 4 and 5 and glycogen synthase kinase 3β are targets for miR-124 and that intrahippocampal infusion of a selective histone deacetylase 4/5 inhibitor or glycogen synthase kinase 3 inhibitor has antidepressant-like actions on behavior. These data suggest that modulation of hippocampal miR-124 pathways may have potential antidepressant effects.

Not only miRNAs but also lncRNAs are involved in negative feedback regulation of the HPA axis. Kino et al. have reported that Gas5, a lncRNA, interacts with glucocorticoid receptors that are activated by the ligand glucocorticoid, interferes with association with the glucocorticoid responsible element (GRE) on DNA, and, as a result, inhibits its transcriptional activity [62]. Thus, Gas5 lncRNA may function as a decoy in glucocorticoid feedback regulation.

We previously identified three miRNAs (miR-449a, miR-34a, and miR-34c) that are predicted to bind to the CRF receptor 3′-untranslated region based on a database search. We reported that expression of miR-449a, but not miR-34a or miR-34c, is increased in the anterior pituitary of restraint-stressed rats [63]. Overexpression of miR-449a suppresses CRF receptor expression, and conversely, knockdown of miR-449a attenuates dexamethasone-induced suppression of CRF receptor expression in primary cultured rat anterior pituitary cells. Thus, stress downregulates the expressions of CRF receptor through glucocorticoids, and miR-449a contributes to the glucocorticoid-induced downregulation of CRF receptor expression in the anterior pituitary. However, we did not find a glucocorticoid-responsive element near the miR-449a coding region; therefore, further studies are needed to clarify the details of the intracellular signaling pathways leading to a glucocorticoid-induced increase in miR-449a expression.

We previously reported that the miRNAs have sequences capable of binding to the glucocorticoid receptor 3′-untranslated region and that overexpression of those miRNAs decreases glucocorticoid receptor expression [64]. As miR-142-3p expression in the hypothalamus and hippocampus was increased by dexamethasone, and the restraint-induced increase in miR-142-3p was blocked by adrenalectomy, miR-142-3p expression appears to be upregulated by glucocorticoids. Therefore, our results suggest that stress-induced miR-142-3p expression in the hypothalamus and hippocampus through corticosterone decreases glucocorticoid receptor expression in these tissues in normal rats.

Although many aspects of the miRNA-related mechanism are unknown, the expression of miRNAs and lncRNAs is altered in response to stress and is definitely involved in the stress response.

7 miRNAs and lncRNAs in DOHaD-Induced Neuropsychiatric Diseases

A few reports have clarified the involvement of miRNAs or lncRNAs in the mechanism by which malnutrition during pregnancy causes neuropsychiatric diseases in children, and several papers have reported changes in miRNA expression in cases of intrauterine growth retardation or fetal growth retardation [65, 66].

The concept of the DOHaD theory is "The inappropriate environment in fertilization, embryos, and infancy interacts with the postnatal environment, which leads to susceptibility to diseases caused by epigenomic changes." Therefore, not only undernutrition in embryos but also the postnatal environment is important. In fact, the environment present during a child's growth influences the risk for disease in the future. Nishi et al. reviewed that maternal separation stress causes not only abnormalities in the HPA axis and stress-related corticosterone levels but also neuropsychiatric disease-like phenotypes. Adult mice that had been subjected to maternal separation stress as pups showed abnormal gene expression in the brain and behavior such as depression, anxiety, or eating disorders [67]. Zucchi et al. reported that prenatal stress upregulates miR-323 and miR-98, which may alter inflammatory responses in the brain and that prenatal stress upregulates miR-219, which targets the gene Dazap1, a putative marker of schizophrenia and bipolar affective disorder in humans [68]. Repressor element-1 silencing transcription (REST) factor, also known as "neuron-restrictive silencing factor," is a transcription factor that binds to the repressor element-1 (RE-1) conserved consensus sequence. Uchida et al. reported that REST is significantly increased in rats exposed to maternal separation stress. Compared to controls, these rats also show a greater HPA axis response to stress and depression-like behavior after chronic stress. Furthermore, expression of pre-miR132, pre-miR-124-1, pre-miR-9-1, pre-miR-9-3, pre-miR-212, and pre-miR-29a, a REST regulating microRNAs, is increased in maternal separation rats [69]. These results suggest that activation of a REST-mediated gene network in the prefrontal cortex at an early stage of postnatal development may contribute to mood and anxiety disorders in response to chronic stressful life events during adulthood. Moss et al. demonstrated postnatal effects of single and repeated glucocorticoid injections during late gestation [70]. They reported that repeated maternal injections of synthetic glucocorticoids prolong gestation, reduce weight at birth and at 3 months of age, are associated with low arterial pressure at 3 months of age, and alter glucose metabolism. Another report examined the influence of abnormalities of the next generation. Mansell et al. reported that maternal anxiety in pregnancy is associated with decreased DNA methylation in the progeny [71]. These results suggest a link between poor maternal mental health and adverse birth outcomes.

We previously showed that corticosterone-induced miR-449a expression in the anterior pituitary is reduced and that inhibition of miR-449a consequently impairs the downregulation of CRF receptor expression during restraint stress in low birth weight rats [72]. We also examined pituitary Gas5 lncRNA expression because Gas5 lncRNA inhibits glucocorticoid receptor binding to the GRE and blocks glucocorticoid-induced gene expression, as described above [62]. We observed that the expression of Gas5 lncRNA was significantly higher in low birth weight rats than in control rats. Overexpression of Gas5 lncRNA likely blocks corticosterone action, thereby causing suppression of miR-449a expression in low birth weight rats despite prolonged elevation of glucocorticoid receptor mRNA expression in the anterior pituitary (Fig. 1). The elevated expression of Gas5 lncRNA may be an obstacle to glucocorticoid receptor autoregulation in the anterior pituitary of low birth weight rats. Thus, miRNAs and lncRNAs related to stress hormone regulation are involved in the development of stress-related diseases. However, the details of the mechanism by which expression of miRNAs or lncRNAs is regulated and the role of profiled miRNAs remain to be fully investigated.

We hypothesize that such impaired corticosterone negative feedback may increase the risk of mental illness in low birth weight rats. As mentioned in the concept of the DOHaD theory, "Disease onset is caused by a mismatch between embryonic memory and the postnatal environment." We have been exploring the possibility that higher glucocorticoids for a long period of time in low birth weight rats may cause neuropsychiatric diseases. The levels of stress-related hormones are associated with the risk of developing metabolic diseases. For example, Murphy et al. reported that treatment with corticosteroid synthesis inhibitor attenuates the effects of diet-induced obesity in female rats exposed to early-life stress [73]. Their results suggest that exposure to stress hormones during early life could be a key event in enhancing diet-induced obesity and metabolic diseases. Furthermore, the nutritional environment after birth changes or enhances the risk of developing disease. For example, Strata et al. reported that diet does not affect the anxiety-like behavior of naturally conceived mice, whereas diet does affect the behavior of mice conceived via in vitro fertilization [74]. Although neither Murphy et al. nor Strata et al. showed data for miRNAs/lncRNAs, we are very interested in expression changes in miRNAs/lncRNAs after dietary intervention.

8 Future Prospects

Clearly, miRNAs and lncRNAs are widely involved in the pathophysiology of the stress response, stress-related neuropsychiatric diseases, and DOHaD. miRNAs not only regulate gene expression in cells but also present in vesi-

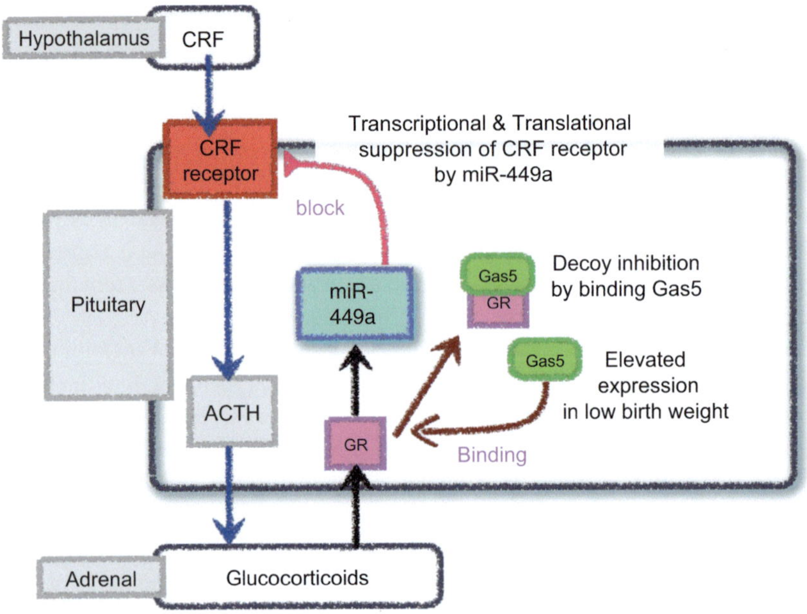

Fig. 1 Noncoding RNA-involved CRF receptor regulation that is potentially associated with stress-related neuropsychiatric disease in low birth weight offspring. In normal rats, CRF stimulates ACTH secretion, which leads to glucocorticoid production in adrenal gland. Glucocorticoids downregulate CRF receptor expression via miR-449a. However, in low birth weight offspring delivered from malnourished dams, Gas 5 rises, lowering the expression of miR-449a which downregulates CRF receptor expression, resulting in a prolonged elevation of plasma corticosterone levels

cles called exosomes and are released from of the cell, acting on other tissues and cells like in an endocrine and paracrine manner. We previously reported that restraint significantly increases serum exosomal miR-449a in control rats but not in low birth weight rats. The sequences of miR-449a and its binding region of CRF receptor 3′-untranslated region are 100% homologous between humans and rats. Quantification of serum exosomal miR-449a may be a useful for evaluating dysregulation of the HPA axis in stressed in humans [64].

Diagnosis and prognosis of intrauterine growth retardation may be made possible by examining the miRNA profile in the maternal blood [75]. We expected that these fields of research will progress and that the findings can be applied to future risk evaluations as suggested by the DOHaD theory to allow early diagnosis. We look forward to contributing to the establishment of preemptive medicine aimed at early diagnosis and intervention by using miRNAs and lncRNAs.

Disclosure Statement The authors have nothing to disclose.

References

1. Newnham JP. The developmental origins of health and disease (DOHaD) – why it is so important to those who work in fetal medicine. Ultrasound Obstet Gynecol. 2007;29(2):121–3. https://doi.org/10.1002/uog.3938.
2. Barker DJ. The origins of the developmental origins theory. J Intern Med. 2007;261(5):412–7. https://doi.org/10.1111/j.1365-2796.2007.01809.x.
3. Gillman MW, Barker D, Bier D, Cagampang F, Challis J, Fall C, Godfrey K, Gluckman P, Hanson M, Kuh D, Nathanielsz P, Nestel P, Thornburg KL. Meeting report on the 3rd International Congress on Developmental Origins of Health and Disease (DOHaD). Pediatr Res. 2007;61(5 Pt 1):625–9. https://doi.org/10.1203/pdr.0b013e3180459fcd.
4. Hanson M. The birth and future health of DOHaD. J Dev Orig Health Dis. 2015;6(5):434–7. https://doi.org/10.1017/S2040174415001129.
5. Fitzhardinge PM, Inwood S. Long-term growth in small-for-date children. Acta Paediatr Scand Suppl. 1989;349:27–33.

6. Karlberg J, Albertsson-Wikland K. Growth in full-term small-for-gestational-age infants: from birth to final height. Pediatr Res. 1995;38(5):733–9. https://doi.org/10.1203/00006450-199511000-00017.
7. Takeuchi A, Yorifuji T, Takahashi K, Nakamura M, Kageyama M, Kubo T, Ogino T, Doi H. Neurodevelopment in full-term small for gestational age infants: A nationwide Japanese population-based study. Brain Dev. 2016;38(6):529–37. https://doi.org/10.1016/j.braindev.2015.12.013.
8. Botellero VL, Skranes J, Bjuland KJ, Lohaugen GC, Haberg AK, Lydersen S, Brubakk AM, Indredavik MS, Martinussen M. Mental health and cerebellar volume during adolescence in very-low-birth-weight infants: a longitudinal study. Child Adolesc Psychiatr Ment Health. 2016;10:6. https://doi.org/10.1186/s13034-016-0093-8.
9. Harvey S, Phillips JG, Rees A, Hall TR. Stress and adrenal function. J Exp Zool. 1984;232(3):633–45. https://doi.org/10.1002/jez.1402320332.
10. Emeric-Sauval E. Corticotropin-releasing factor (CRF) – a review. Psychoneuroendocrinology. 1986;11(3):277–94.
11. Keller-Wood M. Hypothalamic-pituitary – adrenal axis-feedback control. Compr Physiol. 2015;5(3):1161–82. https://doi.org/10.1002/cphy.c140065.
12. de Quervain D, Schwabe L, Roozendaal B. Stress, glucocorticoids and memory: implications for treating fear-related disorders. Nat Rev Neurosci. 2017;18(1):7–19. https://doi.org/10.1038/nrn.2016.155.
13. Oliveira M, Rodrigues AJ, Leao P, Cardona D, Pego JM, Sousa N. The bed nucleus of stria terminalis and the amygdala as targets of antenatal glucocorticoids: implications for fear and anxiety responses. Psychopharmacology. 2012;220(3):443–53. https://doi.org/10.1007/s00213-011-2494-y.
14. Shepard JD, Barron KW, Myers DA. Corticosterone delivery to the amygdala increases corticotropin-releasing factor mRNA in the central amygdaloid nucleus and anxiety-like behavior. Brain Res. 2000;861(2):288–95.
15. Pomara N, Greenberg WM, Branford MD, Doraiswamy PM. Therapeutic implications of HPA axis abnormalities in Alzheimer's disease: review and update. Psychopharmacol Bull. 2003;37(2):120–34.
16. Schelling G. Post-traumatic stress disorder in somatic disease: lessons from critically ill patients. Prog Brain Res. 2008;167:229–37. https://doi.org/10.1016/S0079-6123(07)67016-2.
17. Wosu AC, Valdimarsdottir U, Shields AE, Williams DR, Williams MA. Correlates of cortisol in human hair: implications for epidemiologic studies on health effects of chronic stress. Ann Epidemiol. 2013;23(12):797–811. e792. https://doi.org/10.1016/j.annepidem.2013.09.006.
18. Zhao Y, Ma R, Shen J, Su H, Xing D, Du L. A mouse model of depression induced by repeated corticosterone injections. Eur J Pharmacol. 2008;581(1–2):113–20. https://doi.org/10.1016/j.ejphar.2007.12.005.
19. Huston JP, Komorowski M, de Souza Silva MA, Lamounier-Zepter V, Nikolaus S, Mattern C, Muller CP, Topic B. Chronic corticosterone treatment enhances extinction-induced depression in aged rats. Horm Behav. 2016;86:21–6. https://doi.org/10.1016/j.yhbeh.2016.09.003.
20. Fischer S, Strawbridge R, Vives AH, Cleare AJ. Cortisol as a predictor of psychological therapy response in depressive disorders: systematic review and meta-analysis. Br J Psychiatry. 2017;210(2):105–9. https://doi.org/10.1192/bjp.bp.115.180653.
21. Paskitti ME, McCreary BJ, Herman JP. Stress regulation of adrenocorticosteroid receptor gene transcription and mRNA expression in rat hippocampus: time-course analysis. Brain Res Mol Brain Res. 2000;80(2):142–52.
22. Ordyan NE, Pivina SG, Rakitskaya VV, Shalyapina VG. The neonatal glucocorticoid treatment-produced long-term changes of the pituitary-adrenal function and brain corticosteroid receptors in rats. Steroids. 2001;66(12):883–8.
23. Riedmann LT, Schwentner R. miRNA, siRNA, piRNA and argonautes: news in small matters. RNA Biol. 2010;7(2):133–9.
24. Lee Y, Kim M, Han J, Yeom KH, Lee S, Baek SH, Kim VN. MicroRNA genes are transcribed by RNA polymerase II. EMBO J. 2004;23(20):4051–60. https://doi.org/10.1038/sj.emboj.7600385.
25. Han J, Lee Y, Yeom KH, Kim YK, Jin H, Kim VN. The Drosha-DGCR8 complex in primary microRNA processing. Genes Dev. 2004;18(24):3016–27. https://doi.org/10.1101/gad.1262504.
26. Chendrimada TP, Gregory RI, Kumaraswamy E, Norman J, Cooch N, Nishikura K, Shiekhattar R. TRBP recruits the dicer complex to Ago2 for microRNA processing and gene silencing. Nature. 2005;436(7051):740–4. https://doi.org/10.1038/nature03868.
27. Wilson RC, Tambe A, Kidwell MA, Noland CL, Schneider CP, Doudna JA. Dicer-TRBP complex formation ensures accurate mammalian microRNA biogenesis. Mol Cell. 2015;57(3):397–407. https://doi.org/10.1016/j.molcel.2014.11.030.
28. Schneeberger M, Altirriba J, Garcia A, Esteban Y, Castano C, Garcia-Lavandeira M, Alvarez CV, Gomis R, Claret M. Deletion of miRNA processing enzyme dicer in POMC-expressing cells leads to pituitary dysfunction, neurodegeneration and development of obesity. Mol Metab. 2012;2(2):74–85. https://doi.org/10.1016/j.molmet.2012.10.001.
29. Nishihara T, Zekri L, Braun JE, Izaurralde E. miRISC recruits decapping factors to miRNA targets to enhance their degradation. Nucleic Acids Res. 2013;41(18):8692–705. https://doi.org/10.1093/nar/gkt619.
30. Lu J, Getz G, Miska EA, Alvarez-Saavedra E, Lamb J, Peck D, Sweet-Cordero A, Ebert BL, Mak RH, Ferrando AA, Downing JR, Jacks T, Horvitz HR,

Golub TR. MicroRNA expression profiles classify human cancers. Nature. 2005;435(7043):834–8. https://doi.org/10.1038/nature03702.

31. Henshall DC, Hamer HM, Pasterkamp RJ, Goldstein DB, Kjems J, Prehn JH, Schorge S, Lamottke K, Rosenow F. MicroRNAs in epilepsy: pathophysiology and clinical utility. Lancet Neurol. 2016;15(13):1368–76. https://doi.org/10.1016/S1474-4422(16)30246-0.
32. Mitra B, Rau TF, Surendran N, Brennan JH, Thaveenthiran P, Sorich E, Fitzgerald MC, Rosenfeld JV, Patel SA. Plasma micro-RNA biomarkers for diagnosis and prognosis after traumatic brain injury: a pilot study. J Clin Neurosci. 2017;38:37. https://doi.org/10.1016/j.jocn.2016.12.009.
33. Li N, Pan X, Zhang J, Ma A, Yang S, Ma J, Xie A. Plasma levels of miR-137 and miR-124 are associated with Parkinson's disease but not with Parkinson's disease with depression. Neurol Sci. 2017;38:761. https://doi.org/10.1007/s10072-017-2841-9.
34. Hunter MP, Ismail N, Zhang X, Aguda BD, Lee EJ, Yu L, Xiao T, Schafer J, Lee ML, Schmittgen TD, Nana-Sinkam SP, Jarjoura D, Marsh CB. Detection of microRNA expression in human peripheral blood microvesicles. PLoS One. 2008;3(11):e3694. https://doi.org/10.1371/journal.pone.0003694.
35. Simpson RJ, Lim JW, Moritz RL, Mathivanan S. Exosomes: proteomic insights and diagnostic potential. Expert Rev Proteomics. 2009;6(3):267–83. https://doi.org/10.1586/epr.09.17.
36. Turchinovich A, Weiz L, Langheinz A, Burwinkel B. Characterization of extracellular circulating microRNA. Nucleic Acids Res. 2011;39(16):7223–33. https://doi.org/10.1093/nar/gkr254.
37. Sevignani C, Calin GA, Siracusa LD, Croce CM. Mammalian microRNAs: a small world for fine-tuning gene expression. Mamm Genome. 2006;17(3):189–202. https://doi.org/10.1007/s00335-005-0066-3.
38. Celic T, Meuth VM, Six I, Massy ZA, Metzinger L. The mir-221/222 cluster is a key player in vascular biology via the fine-tuning of endothelial cell physiology. Curr Vasc Pharmacol. 2017;15(1):40–6.
39. Kugel JF, Goodrich JA. The regulation of mammalian mRNA transcription by lncRNAs: recent discoveries and current concepts. Epigenomics. 2013;5(1):95–102. https://doi.org/10.2217/epi.12.69.
40. Mattick JS. Non-coding RNAs: the architects of eukaryotic complexity. EMBO Rep. 2001;2(11):986–91. https://doi.org/10.1093/embo-reports/kve230.
41. Geisler S, Coller J. RNA in unexpected places: long non-coding RNA functions in diverse cellular contexts. Nat Rev Mol Cell Biol. 2013;14(11):699–712. https://doi.org/10.1038/nrm3679.
42. Vance KW, Ponting CP. Transcriptional regulatory functions of nuclear long noncoding RNAs. Trends Genet. 2014;30(8):348–55. https://doi.org/10.1016/j.tig.2014.06.001.
43. Khorkova O, Hsiao J, Wahlestedt C. Basic biology and therapeutic implications of lncRNA. Adv Drug Deliv Rev. 2015;87:15–24. https://doi.org/10.1016/j.addr.2015.05.012.
44. Hu K, Zhang J, Liang M. LncRNA AK015322 promotes proliferation of spermatogonial stem cell C18-4 by acting as a decoy for microRNA-19b-3p. In Vitro Cell Dev Biol Anim. 2016;53:277. https://doi.org/10.1007/s11626-016-0102-5.
45. Wang KC, Chang HY. Molecular mechanisms of long noncoding RNAs. Mol Cell. 2011;43(6):904–14. https://doi.org/10.1016/j.molcel.2011.08.018.
46. Fatica A, Bozzoni I. Long non-coding RNAs: new players in cell differentiation and development. Nat Rev Genet. 2014;15(1):7–21. https://doi.org/10.1038/nrg3606.
47. Clark BS, Blackshaw S. Long non-coding RNA-dependent transcriptional regulation in neuronal development and disease. Front Genet. 2014;5:164. https://doi.org/10.3389/fgene.2014.00164.
48. Smolle E, Haybaeck J. Non-coding RNAs and lipid metabolism. Int J Mol Sci. 2014;15(8):13494–513. https://doi.org/10.3390/ijms150813494.
49. Bischof C, Krishnan J. Exploiting the hypoxia sensitive non-coding genome for organ-specific physiologic reprogramming. Biochimica et Biophysica Acta. 2016;1863(7 Pt B):1782–90. https://doi.org/10.1016/j.bbamcr.2016.01.024.
50. Rinaldi A, Vincenti S, De Vito F, Bozzoni I, Oliverio A, Presutti C, Fragapane P, Mele A. Stress induces region specific alterations in microRNAs expression in mice. Behav Brain Res. 2010;208(1):265–9. https://doi.org/10.1016/j.bbr.2009.11.012.
51. Mannironi C, Camon J, De Vito F, Biundo A, De Stefano ME, Persiconi I, Bozzoni I, Fragapane P, Mele A, Presutti C. Acute stress alters amygdala microRNA miR-135a and miR-124 expression: inferences for corticosteroid dependent stress response. PLoS One. 2013;8(9):e73385. https://doi.org/10.1371/journal.pone.0073385.
52. Schmidt U, Keck ME, Buell DR. miRNAs and other non-coding RNAs in posttraumatic stress disorder: a systematic review of clinical and animal studies. J Psychiatr Res. 2015;65:1–8. https://doi.org/10.1016/j.jpsychires.2015.03.014.
53. Babenko O, Kovalchuk I, Metz GA. Stress-induced perinatal and transgenerational epigenetic programming of brain development and mental health. Neurosci Biobehav Rev. 2015;48:70–91. https://doi.org/10.1016/j.neubiorev.2014.11.013.
54. Issler O, Haramati S, Paul ED, Maeno H, Navon I, Zwang R, Gil S, Mayberg HS, Dunlop BW, Menke A, Awatramani R, Binder EB, Deneris ES, Lowry CA, Chen A. MicroRNA 135 is essential for chronic stress resiliency, antidepressant efficacy, and intact serotonergic activity. Neuron. 2014;83(2):344–60. https://doi.org/10.1016/j.neuron.2014.05.042.
55. Baudry A, Mouillet-Richard S, Schneider B, Launay JM, Kellermann O. miR-16 targets the serotonin transporter: a new facet for adaptive responses to antidepressants. Science. 2010;329(5998):1537–41. https://doi.org/10.1126/science.1193692.

56. Song MF, Dong JZ, Wang YW, He J, Ju X, Zhang L, Zhang YH, Shi JF, Lv YY. CSF miR-16 is decreased in major depression patients and its neutralization in rats induces depression-like behaviors via a serotonin transmitter system. J Affect Disord. 2015;178:25–31. https://doi.org/10.1016/j.jad.2015.02.022.
57. Cui X, Sun X, Niu W, Kong L, He M, Zhong A, Chen S, Jiang K, Zhang L, Cheng Z. Long non-coding RNA: potential diagnostic and therapeutic biomarker for major depressive disorder. Med Sci Monitor. 2016;22:5240–8.
58. Wang Y, Zhao X, Ju W, Flory M, Zhong J, Jiang S, Wang P, Dong X, Tao X, Chen Q, Shen C, Zhong M, Yu Y, Brown WT, Zhong N. Genome-wide differential expression of synaptic long noncoding RNAs in autism spectrum disorder. Transl Psychiatry. 2015;5:e660. https://doi.org/10.1038/tp.2015.144.
59. Ziats MN, Rennert OM. Aberrant expression of long noncoding RNAs in autistic brain. J Mol Neurosci. 2013;49(3):589–93. https://doi.org/10.1007/s12031-012-9880-8.
60. Volk N, Pape JC, Engel M, Zannas AS, Cattane N, Cattaneo A, Binder EB, Chen A. Amygdalar MicroRNA-15a is essential for coping with chronic stress. Cell Rep. 2016;17(7):1882–91. https://doi.org/10.1016/j.celrep.2016.10.038.
61. Higuchi F, Uchida S, Yamagata H, Abe-Higuchi N, Hobara T, Hara K, Kobayashi A, Shintaku T, Itoh Y, Suzuki T, Watanabe Y. Hippocampal MicroRNA-124 enhances chronic stress resilience in mice. J Neurosci. 2016;36(27):7253–67. https://doi.org/10.1523/JNEUROSCI.0319-16.2016.
62. Kino T, Hurt DE, Ichijo T, Nader N, Chrousos GP. Noncoding RNA gas5 is a growth arrest- and starvation-associated repressor of the glucocorticoid receptor. Sci Signal. 2010;3(107):ra8. https://doi.org/10.1126/scisignal.2000568.
63. Nemoto T, Mano A, Shibasaki T. miR-449a contributes to glucocorticoid-induced CRF-R1 downregulation in the pituitary during stress. Mol Endocrinol. 2013;27(10):1593–602. https://doi.org/10.1210/me.2012-1357.
64. Nemoto T, Kakinuma Y, Shibasaki T. Restraint-induced glucocorticoid receptor downregulation is dysregulated in high fat diet-fed rats likely from impairment of miR-142-3p expression in the hypothalamus and hippocampus. Am J Life Sci. 2015;3(3–2):24–30. https://doi.org/10.11648/j.ajls.s.2015030302.15.
65. Chen PY, Ganguly A, Rubbi L, Orozco LD, Morselli M, Ashraf D, Jaroszewicz A, Feng S, Jacobsen SE, Nakano A, Devaskar SU, Pellegrini M. Intrauterine calorie restriction affects placental DNA methylation and gene expression. Physiol Genomics. 2013;45(14):565–76. https://doi.org/10.1152/physiolgenomics.00034.2013.
66. Huang L, Shen Z, Xu Q, Huang X, Chen Q, Li D. Increased levels of microRNA-424 are associated with the pathogenesis of fetal growth restriction. Placenta. 2013;34(7):624–7. https://doi.org/10.1016/j.placenta.2013.04.009.
67. Nishi M, Horii-Hayashi N, Sasagawa T. Effects of early life adverse experiences on the brain: implications from maternal separation models in rodents. Front Neurosci. 2014;8:166. https://doi.org/10.3389/fnins.2014.00166.
68. Zucchi FC, Yao Y, Ward ID, Ilnytskyy Y, Olson DM, Benzies K, Kovalchuk I, Kovalchuk O, Metz GA. Maternal stress induces epigenetic signatures of psychiatric and neurological diseases in the offspring. PLoS One. 2013;8(2):e56967. https://doi.org/10.1371/journal.pone.0056967.
69. Uchida S, Hara K, Kobayashi A, Funato H, Hobara T, Otsuki K, Yamagata H, McEwen BS, Watanabe Y. Early life stress enhances behavioral vulnerability to stress through the activation of REST4-mediated gene transcription in the medial prefrontal cortex of rodents. J Neurosci. 2010;30(45):15007–18. https://doi.org/10.1523/JNEUROSCI.1436-10.2010.
70. Moss TJ, Sloboda DM, Gurrin LC, Harding R, Challis JR, Newnham JP. Programming effects in sheep of prenatal growth restriction and glucocorticoid exposure. Am J Physiol Regul Integr Comp Physiol. 2001;281(3):R960–70.
71. Mansell T, Novakovic B, Meyer B, Rzehak P, Vuillermin P, Ponsonby AL, Collier F, Burgner D, Saffery R, Ryan J, BISi t. The effects of maternal anxiety during pregnancy on IGF2/H19 methylation in cord blood. Transl Psychiatry. 2016;6:e765. https://doi.org/10.1038/tp.2016.32.
72. Nemoto T, Kakinuma Y, Shibasaki T. Impaired miR449a-induced downregulation of Crhr1 expression in low-birth-weight rats. J Endocrinol. 2015;224(2):195–203. https://doi.org/10.1530/JOE-14-0537.
73. Murphy MO, Herald JB, Wills CT, Unfried SG, Cohn DM, Loria AS. Postnatal treatment with metyrapone attenuates the effects of diet-induced obesity in female rats exposed to early-life stress. Am J Phys Endocrinol Metab. 2017;312(2):E98–E108. https://doi.org/10.1152/ajpendo.00308.2016.
74. Strata F, Giritharan G, Sebastiano FD, Piane LD, Kao CN, Donjacour A, Rinaudo P. Behavior and brain gene expression changes in mice exposed to preimplantation and prenatal stress. Reprod Sci. 2015;22(1):23–30. https://doi.org/10.1177/1933719114557900.
75. Cetin I, Mando C, Calabrese S. Maternal predictors of intrauterine growth restriction. Curr Opin Clin Nutr Metab Care. 2013;16(3):310–9. https://doi.org/10.1097/MCO.0b013e32835e8d9c.

Part III

Transgenerational Mechanism and Its Consequences

Placental Development and Nutritional Environment

Kosuke Taniguchi, Tomoko Kawai, and Kenichiro Hata

Abstract

The placenta is considered to have developed recently in mammalian evolution. While the fundamental function of the placenta, i.e., providing nutrients and oxygen to the fetus and receiving waste products, is the same in all mammals, the morphology of the placenta varies substantially in a species-dependent manner. Therefore, considerable interest exists in understanding placental development and function in mammals from a molecular biological viewpoint. Numerous recent studies have shown that various environmental factors before and during pregnancy, including nutrition, affect placental formation and function and that alterations in placental formation and function can influence the developing fetus and the offspring after birth. To date, the relationship between nutrition and the placenta has been investigated in several species, various model organisms, and humans. In this chapter, we discuss the current knowledge of the placenta and the epigenome and then highlight the effects of nutrition during pregnancy on the placenta and the fetus and on the offspring after birth.

K. Taniguchi (✉) · T. Kawai · K. Hata
Department of Maternal-Fetal Biology, National Research Institute for Child Health and Development, Tokyo, Japan
e-mail: hata-k@ncchd.go.jp

Keywords

Epigenome · Placenta · DNA methylation · Genomic imprinting · Retrotransposons

1 Introduction

All mammals possess a placenta, which plays a critical role during pregnancy. The placenta is known to function in efficiently supplying nutrients, exchanging gasses, and receiving waste products while also clearly separating the mother's body from the fetus [1]. However, morphological and pathological investigation of the placenta as a tissue remains limited, and inferring the functions of the placenta based only on morphology or weight might be challenging [2, 3]. Therefore, elucidation of the functions of the placenta from a molecular biological viewpoint is crucial.

During pregnancy, a fertilized egg becomes implanted in the endometrium on day 6 or 7 in the blastocyst stage, which follows the morula stage. In the early blastocyst stage, an inner cell mass and a trophectoderm are present. The inner cell mass differentiates into the embryo, amnion, yolk sac, and allantois, whereas the trophectoderm forms the chorion, which later becomes the major placental tissue [1]. During delivery, the placenta detaches from the decidual surface and emerges from the mother's body to complete its function.

© Springer Nature Singapore Pte Ltd. 2018

T. Kubota, H. Fukuoka (eds.), *Developmental Origins of Health and Disease (DOHaD)*, Advances in Experimental Medicine and Biology 1012, https://doi.org/10.1007/978-981-10-5526-3_7

The fundamental function of the placenta, i.e., providing nutrients and oxygen to the fetus and receiving waste products, is the same in mammals regardless of the species; by contrast, the morphology of the placenta varies substantially in a species-dependent manner and differs considerably from the morphology of other organs such as the heart, lung, and liver [4].

In recent years, numerous studies have reported that the nutritional status before and during pregnancy affects placental formation and function [5, 6] and that alterations in placental formation and function, in turn, can influence the offspring after birth. To date, nutrition and the placenta and their relationship have been investigated in several species, various model organisms, and humans. In this chapter, we begin by discussing the placenta and the epigenome and then consider nutrition and the placenta. We also summarize the status of this field as well as the latest findings in placental epigenetics.

2 Epigenome in the Placenta

2.1 Placental DNA Methylation and Genomic Imprinting

The definition and details of the epigenome are not discussed in this chapter, other than to note that the epigenome, i.e., epigenetic information, refers to the information that determines genetic function without altering the genome (DNA sequence) and is not inherited through the DNA sequence. The typical molecular events/entities that are involved in the epigenome include histone methylation and acetylation and cytosine methylation of DNA and microRNAs (miRNAs) and long noncoding RNAs. During the early stages of development in mammals in particular, the epigenome is recognized to change dynamically for initialization and reorganization. For example, eggs and sperm feature distinct DNA methylation patterns, where DNA methylation in the sperm-derived genome is rapidly removed immediately after fertilization (cell division-independent active demethylation), but in the egg-derived female genome, methylation is slowly lost (passive demethylation) because the DNA strand newly synthesized in cell division is unmethylated. Thus, parent-derived epigenetic information is deleted through the demethylation of DNA from the point immediately after fertilization until around the time of implantation. After implantation, the required DNA methylation is mediated by DNA methyltransferases as development and differentiation progress to establish developmental stage- and tissue-specific DNA methylation (Fig. 1). Therefore, epigenomic reprogramming occurs during the period after fertilization through the early developmental stage, and this should be considered flexible. Moreover, the change in DNA methylation caused by the in utero environmental load during this period might persist until later in life. The developmental origins of health and disease (DOHaD) theory posits that DNA methylation alterations caused by environmental load during the perinatal period are memorized through epigenomic changes, and this theory is based on the finding that DNA methylation information is stably maintained in daughter cells even after cell division.

Conversely, genomic imprinting sites also exist as a part of regions that escape the aforementioned epigenomic reprograming in a genome. The paternal and maternal methylation information in such regions is not deleted after fertilization and is maintained in somatic cells, which indicates that gene expression is regulated by the information imprinted in the parental generation. Therefore, this phenomenon is called genomic imprinting, a mechanism that has been shown to be essential for development/differentiation of the placenta and the fetus [7], and abnormal regulation of genomic imprinting is considered to be associated with various disorders, including miscarriage [8]. For example, as a result of genomic imprinting, diverse functions are performed by *p*aternally *e*xpressed *g*enes: *Peg10* plays a role in placental formation [9], *Peg11/Rtl1* functions in placenta maintenance [10], and *Peg3* regulates the expression of genes involved in placental development [11]. Furthermore, several other imprinted genes have been reported to be associated with fetal growth

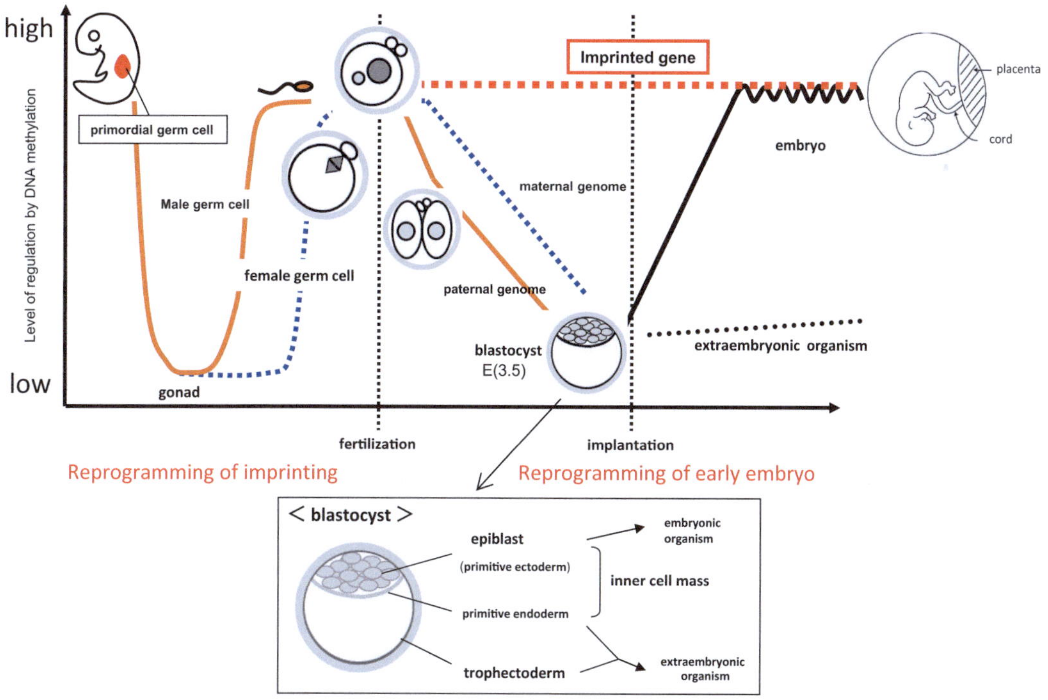

Fig. 1 Epigenetic regulation of tissue-specific genes during development

and placental function [12], and genes that are regulated by genomic imprinting only in the placenta have been identified in humans [13]. However, in contrast to these findings, recent studies have shown that rather than being targeted by the mechanism of genomic imprinting, some of the maternally derived DNA methylation is not deleted after fertilization and is maintained in the placenta, where it is involved in placental development and function [14].

2.2 Other Placental Epigenetic Information

A few studies have reported the functional effects of histone modification and noncoding RNAs in the placenta that are produced by the nutritional environment during pregnancy. Although their functions are incompletely understood, miRNAs are considered to play a critical role in placental development [15]. As noted below, because of the general hypomethylation of DNA in the placenta, this suppressive modification is considered to be switched off from the viewpoint of gene expression, and posttranscriptional regulators such as miRNAs might play a comparatively more important role in the placenta than in other organs.

The behaviors of miRNAs in the placenta have recently been studied in the case of newborns that are small for gestational age (SGA) or exhibit fetal growth restriction (FGR) or macrosomia [16–19], but the function of each miRNA detected thus far remains unclear. The expression of an oncogene, *BCL-2*, and a tumor-suppressor gene, *PTEN*, was further reported to be increased in the villus of patients with recurrent miscarriage, and miRNAs targeting these genes were found to be decreased as compared with the levels in normal study participants (Wang et al. 2016). Moreover, measurement of free placenta-derived miRNAs in maternal blood at the 28th gestational week was suggested to potentially identify late-onset FGR [20].

Abnormal expression of miRNAs in the placenta is speculated to be associated with various diseases, and research in this field might lead to development of novel therapies; thus, future progress in this research area is expected.

2.3 Retrotransposons and Placental Evolution

The discovery that retrotransposon-derived *Peg10* plays a critical role in placental development forms the basis for the following hypotheses. The control of genomic imprinting by differentially methylated regions, a common feature in eutherians, presumably originated from the DNA methylation used as an inactivation mechanism against retrotransposons and foreign sequences in the genome [21]. However, because epigenomic suppression in the placenta is weak, retrotransposons are readily inserted into the genome, and this might drive placental evolution. For example, eutherian-specific *Peg11/Rtl1* is considered to have given rise to the long gestation period that is characteristic of eutherians [10]. Although both *Peg10* and *Peg11/Rtl1* are derived from retrotransposons, these genes have acquired distinct functions that contribute to placental evolution in mammals. In addition to these genes, long terminal repeats (LTRs) are frequently recognized in genes associated with placental formation, such as *Ldoc1*, and several evolutionarily novel genes derived from retrotransposons are suspected to have been acquired [22]. Retrotransposon-derived LTR sequences are also known to be inserted into genomes and to participate in *cis*-acting regulation of gene expression. Moreover, in the human genome, 18 genes derived from retroviral envelope genes (*Env*) have been identified, and, among them, *Syncytins* function in placental formation [23]. Human syncytin-1 (HERV-W), which belongs to the human endogenous retrovirus (HERV) family, is expressed on the cell surface as a glycoprotein and promotes the fusion of placental trophoblast cells into multinucleated syncytial trophoblast cells. Conversely, *Syncytin-2* (*HERV-FRD*) encodes an immunosuppressive domain and thus is suggested to function in protecting the fetus from the mother's immune system. In humans, syncytin is specifically expressed in the placenta and testis. Although the expression level in the testis is considerably lower than that in the placenta, syncytin-1 is expressed in human sperm, and its receptor is expressed in human eggs, and this suggests a potential association of syncytin-1 with the sperm-egg cell fusion at fertilization [24]. Tissue-specific *Syncytin* expression has been reported to be regulated by the epigenetic regulatory mechanism of DNA methylation [25]. The hypomethylation of DNA in the entire placental genome as compared with the methylation level in other tissues (mentioned above) is considered to promote active transcription in the placenta of *HERVs*, which are downregulated by DNA methylation in other tissues. Therefore, retrovirus-derived sequences that are primarily targets of DNA methylation, and are expected to be suppressed owing to this modification, are free from methylation-dependent suppression in the placenta and are involved in placental development and maintenance.

In summary, among somatic tissues, the placenta is controlled by a unique form of epigenetic regulation. The normally regulated status of the placental epigenome is crucial for fetal growth, and changes induced in the epigenome by various environmental factors, including nutrition, might exert diverse effects on the next generation. Elucidation of the status of the placental epigenome might support the DOHaD hypothesis by uncovering the underlying molecular mechanisms.

3 Effects of Nutritional Status During Pregnancy on the Placenta and the Fetus and on the Offspring After Birth

3.1 Effects of Nutritional Status During Pregnancy on the Placenta and the Fetus

Agouti mice are recognized as a mouse model in which the nutritional status during pregnancy can induce changes in an offspring's phenotype, and these changes are mediated by epigenetic regulation and can be clearly detected. In these mice, an endogenous retrovirus-like element, intracisternal A particle (IAP), is inserted 100 kb upstream of the gene *Agouti*, which controls coat color. When IAP, which is readily targeted by the physiological epigenetic mechanism, is highly methylated, IAP transcription is suppressed. However, DNA methylation is a stochastic phenomenon, and the expression of surrounding genes is altered because of IAP in mice when the level of DNA methylation is not particularly high, and this results in the birth of mice featuring yellow fur. Because IAP derepression affects the expression of genes other than *Agouti*, the phenotypes of obesity and diabetes can also be observed. Intake of methyl-group donors, including folate, during the mouse perinatal period (before/during pregnancy and during lactation) has been reported to cause a change in IAP methylation levels, which results in an alteration of coat color in newborns [26], and to also affect the next generation by changing IAP methylation levels in not only somatic cells but also germ cells [27]. These findings indicate that epigenetic regulation in the fetus might be strongly affected by the intrauterine nutritional environment during gestation. Another study reported that in a model organism, poor nutrition in the pregnant mother altered DNA methylation in the germ line of the grandchild's generation through the sperm of the child's generation [28], and other studies have sequentially verified the effects produced on the child's and grandchild's generation by the intrauterine environment. Furthermore, the intrauterine nutritional environment is recognized to affect not only the fetus but also the placenta's morphological and molecular biological phenotypes [29] (Fig. 2).

In humans, newborns who are SGA or large for gestational age (LGA) have been reported to face elevated risks for obesity [30], metabolic syndrome [31, 32], and cardiovascular events [33], possibly because of exposure to an inappropriate intrauterine environment and nongenetic causes such as (primarily) epigenetic factors [34]. Here, we discuss this phenomenon from a nongenetic viewpoint. In maternal hyperglycemia during pregnancy, which is gestational diabetes mellitus (GDM), the fetus/placenta becomes large, and the risk of adverse perinatal outcomes increases [35]. Moreover, regardless of the cause, blood pressure is elevated, and lipid metabolism is abnormal during school-age years in children who were LGA at birth, and they face a future cardio-metabolic risk [36]. These epidemiological findings showed that maternal hyperglycemia during pregnancy represents a potential future disease risk for children. This raises the question of what specific epigenetic changes occur in the placenta of the fetus exposed to maternal hyperglycemia. Among seven imprinted genes associated with fetus/child growth, *mesoderm-specific transcript* (*MEST*) was found to show low methylation rates only in the GDM group. This finding and the considerably lower methylation levels of *MEST* observed in the blood of obese adults together suggest that hyperglycemia during pregnancy might cause a change in DNA methylation in the placenta that is similar to the methylation change observed in metabolic syndrome [37]. Maternal hyperglycemia alters the methylation of genes associated with brown adipose tissue in the placenta [38]. Although the DNA in the placenta is hypomethylated overall (as noted in the preceding section), one study reported that mothers with GDM displayed markedly increased levels of global placental DNA methylation [39]. Another study showed that in mice, a 50% reduction of nutrient intake during pregnancy tended to increase the number of hypomethylated sites in the placenta, and this depended on fetal sex [40]. Thus, hyperglycemia during pregnancy causes epigenetic changes in the placenta, and, furthermore, this status change is similar to that observed in metabolic syndrome in certain cases and might become a cause of the fetal reprograming that leads to future disorders.

Poor maternal nutritional status (limited intake of calories) can result in the birth of SGA newborns and, subsequently, in an increase in their body fat content, leading to obesity in adulthood [41]. In the placenta of SGA newborns, DNA methylation of genes associated with energy homeostasis was reported to be altered [42], which suggests that the maternal environment

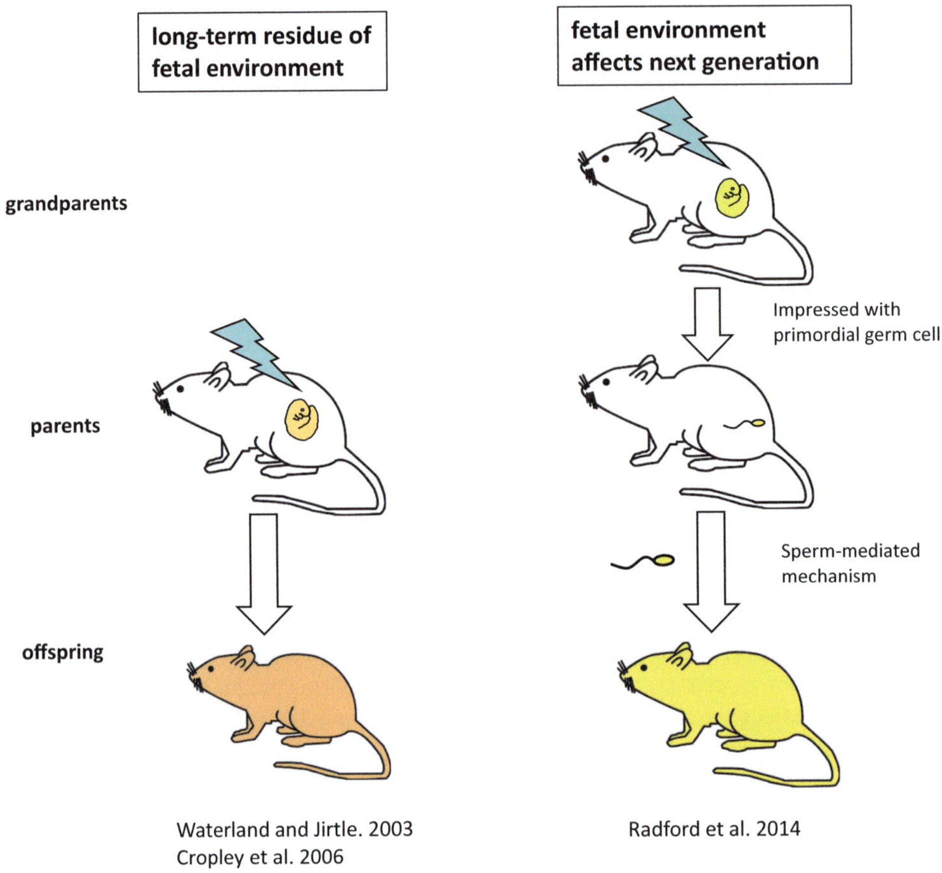

Fig. 2 Alteration of the environmental epigenome

during pregnancy could affect the next generation. Because numerous SGA studies have been conducted, including epidemiological studies, we reserve the discussion of this work for the next chapter.

Folate, a methyl-group donor, is a key nutrient during pregnancy, and therefore, the effect of folate consumption during pregnancy has been extensively investigated. The importance of folate has also been illustrated by a recent study demonstrating that the expression levels of genes associated with folate transport and the levels of the related proteins in the placenta were lower in SGA and LGA newborns than in normal controls [43]. Gene ontology analysis was performed on the genes located in the chromosomal regions where DNA methylation was altered in the placenta of mice that were deficient in folate during pregnancy, and the results showed significant associations of terms including biological adhesion, biological regulation, cell proliferation, development, metabolism, and signaling, in which regulators of decidua formation were observed [44]. Another study tested the effects of adding controlled amounts of folate to two human placental cell lines; while treatment with excess folate led to a decrease in cell viability but an increase in growth rate, exposure to deficient folate amounts resulted in a decrease in both cell viability and invasiveness [45]. Moreover, in the absence of folate intake, decidual angiogenesis is suppressed [46]. Therefore, proper folate intake during pregnancy is necessary for placental development and formation. Furthermore, in the promoter of *methylenetetrahydrofolate reductase* (*MTHFR*), a gene encoding a folate metabolic enzyme, the methylation rate was elevated in the placenta and the maternal peripheral blood of patients with preeclampsia [47], which indicates that folate is associated with diseases diagnosed during pregnancy in humans.

As noted earlier in this chapter, the status of the epigenome in the placenta can be readily altered by the environment. We have focused on this aspect of the epigenome in our research, with the expectation that changes induced in DNA methylation of nonspecific genes by environmental stress will be readily observed in the placenta. In humans, the nutritional environment during pregnancy cannot be arbitrarily controlled, and large individual variations are produced due to distinct genetic backgrounds and living environments; consequently, fewer findings regarding the nutritional conditions and epigenetic regulation during pregnancy in humans have been reported than in the case of animal models. Conversely, as an example of the association between the environment during the human embryonic stage and epigenetic regulation, we identified a phenomenon where the placental epigenome varies depending on maternal body weight gain during pregnancy. Our study demonstrated that even when the birth weight of a fetus is normal, a higher deviation of gestational weight gain from the appropriate value is associated more frequent detection of outliers of DNA methylation in the placenta. This result suggests that improper weight gain during pregnancy increases the frequency with which changes are induced in placental epigenetic regulation [48] (Fig. 3).

3.2 Effects of Placental Formation on the Offspring After Birth

The association between nutritional status during pregnancy and the development of various diseases has become widely recognized as a result of the studies that formed the basis of the Barker hypothesis [49]: the Dutch famine study [50, 51] and epidemiological studies such as the Hertfordshire Cohort Study and the Helsinki Birth Cohort Study [52]. These epidemiological studies provided valuable information such as the effects produced on offspring in extremely unfavorable environments (which cannot be readily replicated experimentally), reflected by the particular times at which the infants were born. However, to be able to implement these findings in society, future studies must demonstrate the association between the inappropriate environments during pregnancy that can develop in the current social/living/medical environment and the future onset of disease.

A study on pregnant women who had experienced the Dutch famine [53] reported that regardless of pregnancy term, placental size was small. However, compared with the size predicted from their placental area, newborns who were in mid-late gestation during the famine were light, and newborns who were in early gestation during the famine, or who were conceived after it had ended, were heavy. This indicates that alterations in the crucial nutritional environment during pregnancy, including preconception, impair the normal processes of placentation and change placental morphology, which in turn influences newborn weight through an unknown mechanism. The placenta's morphological features, such as weight and area, have been reported to be associated with the risk of future cardiovascular disease and hypertension, and these features can therefore serve as potential indices [50, 54]. One study showed that when the mother's height was ≤160 cm, the incidence of hypertension was high in the group featuring a small placental area [54]. Furthermore, maternal diet, weight gain, and nutrition were also shown in an animal study to be associated with placental inflammation/nutrient transport/oxidative stress in addition to placental function and to affect the health of the offspring [6]. In humans, the placenta matures over a period of 40 weeks, but the effects produced during the process of placental formation can affect the offspring even several decades after birth. Moreover, although the finding was limited to males, the Dutch famine study showed that placental area, thickness, and shape (round or not) are associated with hypertension and that when the placental area is higher than a certain value, it might become a risk factor for hypertension [55].

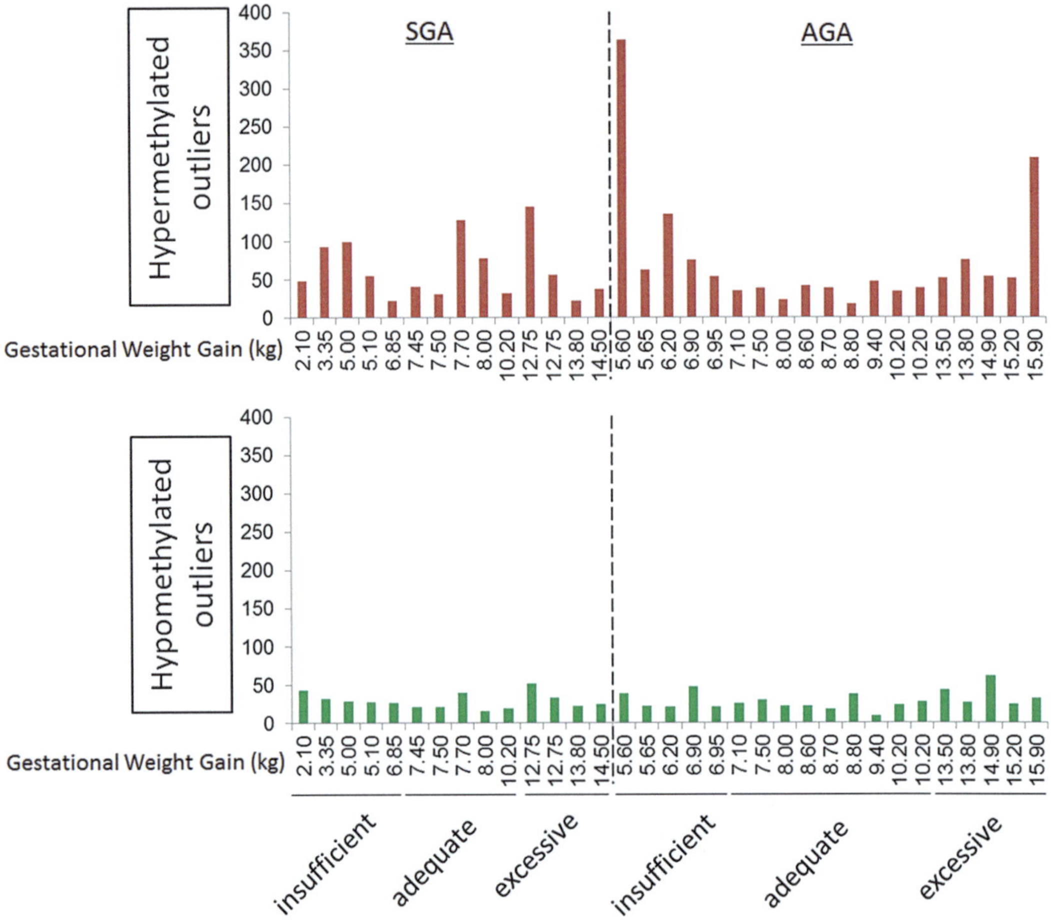

Fig. 3 Inadequate gestational weight gain induces unique changes in placental DNA methylation. *SGA* small for gestational age, *AGA* appropriate for gestational age

With regard to nutrition during pregnancy, placental weight was found to show no association with intake of an unbalanced high-protein (meat) or low-carbohydrate diet during late gestation, but such diets represented a risk factor for future hypertension in the born children, the cause of which was associated with high cortisol levels when the children became adults [56, 57]. In pigs, both deficient and excessive intake of proteins as compared to carbohydrates during pregnancy resulted in alterations in the expression of glucocorticoid-related genes [58] and affected carbohydrate metabolism in the liver [59]. Therefore, nutrition during pregnancy can also alter hormonal balance through the placenta.

In humans, only a few studies based on an epidemiological design have been reported to date on placental epigenetics and long-term prognosis. A study on short-term prognosis showed that certain genes exhibited altered DNA methylation in the placenta and cord blood of an SGA neonate born at full term and that these genes were correlated with body fat at 2 weeks after birth [42]. Another study on the neonatal phenotype associated with the status of the epigenome in the placenta reported that methylation in the placenta of the genes *fragile histidine triad* (*FHIT*) and *ankyrin repeat domain 11* (*ANKRD11*), which are associated with neural development and behavior, was strongly associated with attention capacity in the neonates [60].

4 Conclusion

The maternal status produced before and during pregnancy by various environmental factors, including nutrition, induces epigenetic changes in the placenta and thereby affects the phenotypes of both the fetus and the child after birth. While the details of the specific changes induced in various organs are addressed in other sections, in this chapter, our focus has been on the placenta. The placenta is an organ that is considered to have developed recently in mammalian evolution. As mentioned in this chapter's introduction section, the function of the placenta is similar in all mammals, but the microstructure of the placenta shows large species-dependent differences. Retrotransposon-derived genes, which are maintained in a hypomethylated state in the placenta after differentiation from a fertilized egg and in a hypermethylated state in somatic tissues, play a crucial role in placental development and differentiation; thus, the placenta is a unique organ from an epigenetic viewpoint. However, this feature might cause the epigenetic changes that are induced by environmental factors to persist longer than the embryonic portion, and thus the placenta could emerge as a favorable standard epigenetic organ for inferring the fetal environment.

References

1. Benirschke K, Burton GJ, Baergen RN. Pathology of the human placenta. 6th ed. Berlin/Heidelberg: Springer; 2012.
2. David AL, Jauniaux E. Ultrasound and endocrinological markers of first trimester placentation and subsequent fetal size. Placenta. 2016;40:29–33.
3. Heinonen S, Taipale P, Saarikoski S. Weights of placentae from small-for-gestational age infants revisited. Placenta. 2001;22:399–404.
4. Enders AC, Carter AM. What can comparative studies of placental structure tell us? A review. Placenta. 2004;25(Suppl A):S3–9.
5. Tarrade A, Panchenko P, Junien C, Gabory A. Placental contribution to nutritional programming of health and diseases: epigenetics and sexual dimorphism. J Exp Biol. 2015;218:50–8.
6. Thornburg KL, O'Tierney PF, Louey S. Review: the placenta is a programming agent for cardiovascular disease. Placenta. 2010;31(Suppl):S54–9.
7. Monk D. Genomic imprinting in the human placenta. Am J Obstet Gynecol. 2015;213:S152–62.
8. Hanna CW, McFadden DE, Robinson WP. DNA methylation profiling of placental villi from karyotypically normal miscarriage and recurrent miscarriage. Am J Pathol. 2013;182:2276–84.
9. Ono R, Nakamura K, Inoue K, Naruse M, Usami T, Wakisaka-Saito N, et al. Deletion of Peg10, an imprinted gene acquired from a retrotransposon, causes early embryonic lethality. Nat Genet. 2006;38:101–6.
10. Sekita Y, Wagatsuma H, Nakamura K, Ono R, Kagami M, Wakisaka N, et al. Role of retrotransposon-derived imprinted gene, Rtl1, in the fetomaternal interface of mouse placenta. Nat Genet. 2008;40:243–8.
11. Kim J, Frey WD, He H, Kim H, Ekram MB, Bakshi A, et al. Peg3 mutational effects on reproduction and placenta-specific gene families. PLoS One. 2013;8:e83359.
12. Moore GE, Ishida M, Demetriou C, Al-Olabi L, Leon LJ, Thomas AC, et al. The role and interaction of imprinted genes in human fetal growth. Philos Trans R Soc Lond Ser B Biol Sci. 2015;370:20140074.
13. Court F, Tayama C, Romanelli V, Martin-Trujillo A, Iglesias-Platas I, Okamura K, et al. Genome-wide parent-of-origin DNA methylation analysis reveals the intricacies of human imprinting and suggests a germline methylation-independent mechanism of establishment. Genome Res. 2014;24:554–69.
14. Sanchez-Delgado M, Court F, Vidal E, Medrano J, Monteagudo-Sanchez A, Martin-Trujillo A, et al. Human oocyte-derived methylation differences persist in the placenta revealing widespread transient imprinting. PLoS Genet. 2016;12:e1006427.
15. Gu Y, Sun J, Groome LJ, Wang Y. Differential miRNA expression profiles between the first and third trimester human placentas. Am J Physiol Endocrinol Metab. 2013;304:E836–43.
16. Higashijima A, Miura K, Mishima H, Kinoshita A, Jo O, Abe S, et al. Characterization of placenta-specific microRNAs in fetal growth restriction pregnancy. Prenat Diagn. 2013;33:214–22.
17. Tang Q, Wu W, Xu X, Huang L, Gao Q, Chen H, et al. miR-141 contributes to fetal growth restriction by regulating PLAG1 expression. PLoS One. 2013;8:e58737.
18. Li J, Song L, Zhou L, Wu J, Sheng C, Chen H, et al. A MicroRNA signature in gestational diabetes mellitus associated with risk of Macrosomia. Cell Physiol Biochem. 2015;37:243–52.
19. Li J, Chen L, Tang Q, Wu W, Gu H, Liu L, et al. The role, mechanism and potentially novel biomarker of microRNA-17-92 cluster in macrosomia. Sci Rep. 2015;5:17212.

20. Whitehead CL, McNamara H, Walker SP, Alexiadis M, Fuller PJ, Vickers DK, et al. Identifying late-onset fetal growth restriction by measuring circulating placental RNA in the maternal blood at 28 weeks' gestation. Am J Obstet Gynecol. 2016;214:521 e1-8.
21. Suzuki S, Ono R, Narita T, Pask AJ, Shaw G, Wang C, et al. Retrotransposon silencing by DNA methylation can drive mammalian genomic imprinting. PLoS Genet. 2007;3:e55.
22. Naruse M, Ono R, Irie M, Nakamura K, Furuse T, Hino T, et al. Sirh7/Ldoc1 knockout mice exhibit placental P4 overproduction and delayed parturition. Development. 2014;141:4763–71.
23. Cornelis G, Vernochet C, Carradec Q, Souquere S, Mulot B, Catzeflis F, et al. Retroviral envelope gene captures and syncytin exaptation for placentation in marsupials. Proc Natl Acad Sci U S A. 2015;112:E487–96.
24. Bjerregaard B, Lemmen JG, Petersen MR, Ostrup E, Iversen LH, Almstrup K, et al. Syncytin-1 and its receptor is present in human gametes. J Assist Reprod Genet. 2014;31:533–9.
25. Nelissen EC, van Montfoort AP, Dumoulin JC, Evers JL. Epigenetics and the placenta. Hum Reprod Update. 2011;17:397–417.
26. Waterland RA, Jirtle RL. Transposable elements: targets for early nutritional effects on epigenetic gene regulation. Mol Cell Biol. 2003;23:5293–300.
27. Cropley JE, Suter CM, Beckman KB, Martin DI. Germ-line epigenetic modification of the murine a vy allele by nutritional supplementation. Proc Natl Acad Sci U S A. 2006;103:17308–12.
28. Radford EJ, Ito M, Shi H, Corish JA, Yamazawa K, Isganaitis E, et al. In utero effects. In utero undernourishment perturbs the adult sperm methylome and intergenerational metabolism. Science. 2014;1255903:345.
29. Coan PM, Vaughan OR, Sekita Y, Finn SL, Burton GJ, Constancia M, et al. Adaptations in placental phenotype support fetal growth during undernutrition of pregnant mice. J Physiol. 2010;588:527–38.
30. Gallo P, Cioffi L, Limauro R, Farris E, Bianco V, Sassi R, et al. SGA children in pediatric primary care: what is the best choice, large or small? A 10-year prospective longitudinal study. Glob Pediatr Health. 2016;3:2333794X16659993.
31. Chiavaroli V, Giannini C, D'Adamo E, de Giorgis T, Chiarelli F, Mohn A. Insulin resistance and oxidative stress in children born small and large for gestational age. Pediatrics. 2009;124:695–702.
32. Matta J, Carette C, Levy Marchal C, Bertrand J, Petera M, Zins M, et al. Weight for gestational age and metabolically healthy obesity in adults from the Haguenau cohort. BMJ Open. 2016;6:e011367.
33. Mericq V, Martinez-Aguayo A, Uauy R, Iniguez G, Van der Steen M, Hokken-Koelega A. Long-term metabolic risk among children born premature or small for gestational age. Nat Rev Endocrinol. 2017;13:50–62.
34. Han DY, Murphy R, Morgan AR, Lam WJ, Thompson JM, Wall CR, et al. Reduced genetic influence on childhood obesity in small for gestational age children. BMC Med Genet. 2013;14:10.
35. Farrar D, Simmonds M, Bryant M, Sheldon TA, Tuffnell D, Golder S, et al. Hyperglycaemia and risk of adverse perinatal outcomes: systematic review and meta-analysis. BMJ. 2016;354:i4694.
36. Lin XH, Wu DD, Gao L, Zhang JY, Pan HT, Wang H, et al. Altered DNA methylation in neonates born large-for-gestational-age is associated with cardiometabolic risk in children. Oncotarget. 2016;7:86511.
37. El Hajj N, Pliushch G, Schneider E, Dittrich M, Muller T, Korenkov M, et al. Metabolic programming of MEST DNA methylation by intrauterine exposure to gestational diabetes mellitus. Diabetes. 2013;62:1320–8.
38. Cote S, Gagne-Ouellet V, Guay SP, Allard C, Houde AA, Perron P, et al. PPARGC1alpha gene DNA methylation variations in human placenta mediate the link between maternal hyperglycemia and leptin levels in newborns. Clin Epigenetics. 2016;8:72.
39. Reichetzeder C, Dwi Putra SE, Pfab T, Slowinski T, Neuber C, Kleuser B, et al. Increased global placental DNA methylation levels are associated with gestational diabetes. Clin Epigenetics. 2016;8:82.
40. Chen PY, Ganguly A, Rubbi L, Orozco LD, Morselli M, Ashraf D, et al. Intrauterine calorie restriction affects placental DNA methylation and gene expression. Physiol Genomics. 2013;45:565–76.
41. Yura S, Itoh H, Sagawa N, Yamamoto H, Masuzaki H, Nakao K, et al. Role of premature leptin surge in obesity resulting from intrauterine undernutrition. Cell Metab. 2005;1:371–8.
42. Diaz M, Garcia C, Sebastiani G, de Zegher F, Lopez-Bermejo A, Ibanez L. Placental and cord blood methylation of genes involved in energy homeostasis: association with fetal growth and neonatal body composition. Diabetes. 2017;66:779–84.
43. Caviedes L, Iniguez G, Hidalgo P, Castro JJ, Castano E, Llanos M, et al. Relationship between folate transporters expression in human placentas at term and birth weights. Placenta. 2016;38:24–8.
44. Geng Y, Gao R, Chen X, Liu X, Liao X, Li Y, et al. Folate deficiency impairs decidualization and alters methylation patterns of the genome in mice. Mol Hum Reprod. 2015;21:844–56.
45. Ahmed T, Fellus I, Gaudet J, MacFarlane AJ, Fontaine-Bisson B, Bainbridge SA. Effect of folic acid on human trophoblast health and function in vitro. Placenta. 2016;37:7–15.
46. Li Y, Gao R, Liu X, Chen X, Liao X, Geng Y, et al. Folate deficiency could restrain decidual angiogenesis in pregnant mice. Forum Nutr. 2015;7:6425–45.
47. Ge J, Wang J, Zhang F, Diao B, Song ZF, Shan LL, et al. Correlation between MTHFR gene methylation and pre-eclampsia, and its clinical significance. Genet Mol Res. 2015;14:8021–8.
48. Kawai T, Yamada T, Abe K, Okamura K, Kamura H, Akaishi R, et al. Increased epigenetic alterations at the promoters of transcriptional regulators following

inadequate maternal gestational weight gain. Sci Rep. 2015;5:14224.
49. Barker DJ, Osmond C. Infant mortality, childhood nutrition, and ischaemic heart disease in England and Wales. Lancet. 1986;1:1077–81.
50. Painter RC, de Rooij SR, Bossuyt PM, Simmers TA, Osmond C, Barker DJ, et al. Early onset of coronary artery disease after prenatal exposure to the Dutch famine. Am J Clin Nutr. 2006;84:322–7. quiz 466-7
51. Ravelli AC, van Der Meulen JH, Osmond C, Barker DJ, Bleker OP. Obesity at the age of 50 y in men and women exposed to famine prenatally. Am J Clin Nutr. 1999;70:811–6.
52. Syddall HE, Sayer AA, Simmonds SJ, Osmond C, Cox V, Dennison EM, et al. Birth weight, infant weight gain, and cause-specific mortality: the Hertfordshire Cohort Study. Am J Epidemiol. 2005;161:1074–80.
53. Roseboom TJ, Painter RC, de Rooij SR, van Abeelen AF, Veenendaal MV, Osmond C, et al. Effects of famine on placental size and efficiency. Placenta. 2011;32:395–9.
54. Barker DJ, Thornburg KL, Osmond C, Kajantie E, Eriksson JG. The surface area of the placenta and hypertension in the offspring in later life. Int J Dev Biol. 2010;54:525–30.
55. van Abeelen AF, de Rooij SR, Osmond C, Painter RC, Veenendaal MV, Bossuyt PM, et al. The sex-specific effects of famine on the association between placental size and later hypertension. Placenta. 2011;32:694–8.
56. Reynolds RM, Godfrey KM, Barker M, Osmond C, Phillips DI. Stress responsiveness in adult life: influence of mother's diet in late pregnancy. J Clin Endocrinol Metab. 2007;92:2208–10.
57. Herrick K, Phillips DI, Haselden S, Shiell AW, Campbell-Brown M, Godfrey KM. Maternal consumption of a high-meat, low-carbohydrate diet in late pregnancy: relation to adult cortisol concentrations in the offspring. J Clin Endocrinol Metab. 2003;88:3554–60.
58. Kanitz E, Otten W, Tuchscherer M, Grabner M, Brussow KP, Rehfeldt C, et al. High and low protein ratio carbohydrate dietary ratios during gestation alter maternal-fetal cortisol regulation in pigs. PLoS One. 2012;7:e52748.
59. Metges CC, Gors S, Lang IS, Hammon HM, Brussow KP, Weitzel JM, et al. Low and high dietary protein:carbohydrate ratios during pregnancy affect materno-fetal glucose metabolism in pigs. J Nutr. 2014;144:155–63.
60. Paquette AG, Houseman EA, Green BB, Lesseur C, Armstrong DA, Lester B, et al. Regions of variable DNA methylation in human placenta associated with newborn neurobehavior. Epigenetics. 2016;11:603–13.

Risk of Neurodevelopmental Disease by Paternal Aging: A Possible Influence of Epigenetic Alteration in Sperm

Ryuichi Kimura, Kaichi Yoshizaki, and Noriko Osumi

Abstract

Sincethe theory of DOHaD has been thrown in the spotlight, most attention has focused on environmental effects of the uterus on developing embryos/fetuses. However, the ontogenesis traces back to gametogenesis. Compared to oogenesis, spermatogenesis goes through far more cell divisions and is therefore more prone to genetic variation and epigenetic alterations. This article will mainly discuss recent findings about the effects of the advanced paternal age on the next generation, in relation to the onset of psychiatric disorders such as autism spectrum disorder. We would like to advocate for further exploration on the DOHaD theory in a wider view.

Keywords

Paternal aging · Sperm epigenetics · Neurodevelopmental disorder

R. Kimura · K. Yoshizaki · N. Osumi (✉)
Graduate School of Medicine, Tohoku University, Sendai, Japan
e-mail: osumi@med.tohoku.ac.jp

1 Introduction

During World War II, transportation of supplies to the Dutch stopped, which brought about a severe famine, reducing nutritious intake below a few hundred calories. A large-scale survey of 300,000 people was conducted to find out what kinds of diseases were developing in adulthood for those who experienced the famine in their mother's womb. The initial finding reported by Ravelli et al. in 1976 in the *New England Journal of Medicine* was about increase in obesity [1], and later, an increase in onset of diabetes and heart diseases was also reported [2, 3]. On the other hand, in a 1986 report in Lancet, Barker and Osmond in the UK found that in areas with high neonatal mortality rates from 1921 to 1925, there was also a high rate of death from cardiovascular problems observed from 1969 to 1978 [4]. These data formed the basis of theories known as "Developmental Origin of Health and Disease (DOHaD) hypothesis," "Hypothesis on Fetal Origins of Lifestyle-Related Diseases," and "Fetal Reprogramming Hypothesis" or taking the name of its proponent – "Barker Hypothesis,"

In the fields of teratology and congenital abnormality, it is widely known that in addition to genetic causes, a variety of birth defects appear due to various environmental factors. In DOHaD hypothesis, even if the environmental effects at the developing stage are not apparent at birth, it has an effect on the onset of diseases later in life.

© Springer Nature Singapore Pte Ltd. 2018

T. Kubota, H. Fukuoka (eds.), *Developmental Origins of Health and Disease (DOHaD)*, Advances in Experimental Medicine and Biology 1012, https://doi.org/10.1007/978-981-10-5526-3_8

The difference seems to be the point of focus in the timeline of disease onset.

DOHaD initially advocated correlation to the risk of metabolic diseases or lifestyle-related diseases, but Susser and Lin who studied the aforementioned Dutch famine cohort reported an approximately twofold increase in the onset of schizophrenia [5]. A similar survey was conducted to study the effects of a large-scale famine that took place in China between 1959 and 1961 [6]. From these epidemiological data, it was thought that certain developmental impairment in the brain led to the onset of postpubertal schizophrenia (the neurodevelopmental theory). Presently in Japan, there is a steady rise of infants born with low birth weight below 2500 g. According to the statistics from the Ministry of Health, Labour and Welfare, the percentage of low birth weight infants has risen up to 9.6% of the total number of live births [7]. Since the brain is the organ with the highest energetic requirement in the body, it is not surprising that low birth weight can affect onset of mental disorders.

As mentioned above, DOHaD theory mainly focuses on the "intrauterine environment" with regard to "developmental effects," but can we leave the prenatal effects from the paternal side out of consideration? Biological effects from the paternal side are introduced into the next-generation individuals via the sperm cell and influence on the developmental process. In this article, we will discuss the increasing amount of data gathered about the effect of paternal aging on the onset of mental disorders in the next generation and the possible mechanism of epigenetic alteration.

2 Risk of Neurodevelopmental Disorders Due to Paternal Aging

In recent years, there is an increased prevalence of autism spectrum disorder (ASD). According to Weintraub's article in Nature in 2011, even though the prevalence was 1 out of 5000 in 1975, this would increase to 1 out of 110 by 2009 [8]. In America's Centers for Disease Control and Prevention reports, the prevalence rate has been further increased: 1 out of 88 in 2012 and 1 out of 68 in 2014. It is admitted that nonbiological factors such as better defined diagnostic criteria for ASD and its greater awareness in society within the past 30–40 years are certainly behind the rapid increase. For instance, there are quite a number of cases where children who would previously have been diagnosed with intellectual disability now fit the diagnostic criteria for ASD. Also, tendency of parents and teachers who are suspicious on behavior of their children/students consult specialists can lead to an increase in the rate of detecting ASD.

From classic twin studies, due to the clearly high concordance rate for ASD, the search for responsible genes had been moving ahead quickly. However, few cases fit with classical Mendelian genetics, and therefore multiple factors are considered to be involved. Nonetheless, it is hard to assume that the increased prevalence of ASD is due to the rapid increase in the number of risk gene carriers within the population, in a period of just 30–40 years. It is therefore meaningful to search for some kind of biological factors such as genetic and epigenetic mechanisms.

The age of the parents is one of such biological factors. There is a rise in age of marriage in all developed countries, and subsequently the age of parents who have their babies is also shifting upward. This trend is also driving the use of assisted reproductive technology. Reichenberg et al. looked at the birth cohort data collected in Israel and classified the frequency of ASD in a group of 132,161 individuals according to the father's age [9]. Here, given the risk for ASD in children who were delivered when the father's age was below 29 years old was 1, they discovered that the risk was 5.65 times higher when the father was in his 40s. Reichenberg's group then published the meta-analysis of the survey conducted in Denmark, Norway, Sweden, Australia, and Israel [10]. A total of 5,766,794 children born between 1985 and 2004 were tracked from 2004 to 2009, and 30,902 children were found to be autistic. In this group, an estimate of the effects of the parent's age against the risk of developing ASD showed that when comparing mothers in their 40s with mothers in their 20s, the

Table 1 Paternal Age and Risk of ASD

Age of Father	RR (95%CI)
< 20	1.08 (0.92–1.27)
20–29	1
30–39	1.05 (1.02–1.08)
40–49	1.28 (1.22–1.34)
> 50	1.66 (1.49–1.85)

RR relative risk, *CI* confidence interval
Sandin et al. [10], modified

Table 2 Paternal Age and Risk of Schizophrenia

Age of Father	RR (95%CI)
20–24	1
25–29	1.14 (0.84–1.53)
30–34	1.42 (1.03–1.96)
35–39	1.64 (1.13–2.38)
40–44	1.73 (1.11–2.70)
45–49	2.02 (1.17–3.51)
50–54	2.96 (1.60–5.47)

RR relative risk, *CI* confidence interval
Malaspina et al. [12], modified

risk of a child developing ASD is 1.15 times higher. Whereas when comparing fathers in their 50s and above with fathers in their 20s, it was found that the child's risk of developing ASD rises up to 1.66 times higher (Table 1). Therefore, we can regard the advanced paternal age as causing more vulnerability for children developing ASD than the advanced maternal age. Such an influence of paternal aging is also reported in schizophrenia (Table 2) [11, 12]. More large-scale meta-analysis is necessary to learn more about the effects of the advanced paternal age with regard to other mental disorders.

3 The Effect of Advanced Paternal Age on De Novo Mutation and Methylation of the Sperm

The first possible cause that can explain why there is an increased risk of ASD due to paternal aging is the rise in de novo mutations. In genetic analyses of ASD, it was reported that de novo mutations, rather than those inherited from the parents, seemed to affect more as ASD risk of children. For example, in 2012, three papers on exome sequencing were published side-by-side in Nature [13–15]. Among them, O'Roak et al. found that de novo mutations of sporadic autistic children originated from father are four times as much and further reported that the advanced paternal age led to an increase in the de novo mutations. Also, in a later study by Kong et al. with an average father's age of 29.7, the average de novo mutation rate is estimated to be 1.20×10^{-8} per nucleotide [16]. The number of mutations increases by about 2 bases per year, with mutations estimated to double every 16.5 years of age.

The reason why most of such de novo mutations traced back to the father is that spermatogenesis has a higher risk of genetic mutation than oogenesis. Spermatogonia, which are sperm stem cells, replicate themselves over one's lifetime and generate spermatocytes, i.e., sperm progenitors that undergo meiosis to produce differentiated spermatozoa. Because many cell divisions are taking place during this process, it may lead a risk of error in the DNA replication. Other possibilities are that some kind of DNA damage starts in spermatogonia as one gets older or DNA repair mechanism deteriorates with aging. On the other hand, for oogenesis, hundreds of oocytes enter a resting phase halfway through meiosis during the fetal stage in the ovary. Following puberty, ovulation occurs according to the menstrual cycle, and the meiosis is completed only after fertilization. It is therefore difficult for de novo mutation to occur.

Now, what about epigenetic change? In 2014, an analysis on age-associated alterations in sperm DNA methylation was published [17]. Sperm samples were collected twice from 17 fertile donors with an interval of 9–19 years from the first sperm collection, and global methylation of sperm was compared among the cohorts. The results identified 139 hypomethylated regions and 8 hypermethylated regions with aging. These age-associated alterations in sperm DNA methylation were further confirmed in the following study. In a 2015 paper, an analysis of 12 sperm samples showed selected CpG sites strongly showing correlation with aging, and additional 68 semen samples obtained from individuals aged 20–73 years revealed 3 CpG sites showing a high correlation with advanced ages [18]. Although the mechanism has been unclear as well as the case of de novo mutations, it is evident that alteration of sperm DNA methylation can also develop by paternal aging.

Research about causal relationship between alterations in sperm DNA methylation and the ASD risk for children has only just begun. In a previous study, alterations in DNA methylation were found in the promoter or gene body of 117 genes including those associated with schizophrenia and bipolar disorder [17]. Also, the US Early Autism Risk Longitudinal Investigation, a longitudinal study of families that already have a child diagnosed with ASD and future offspring who are considered as higher-risk cohort of ASD, gathers data about the factors in the future offspring's ASD risk prediction. In recent results of this study, correlation between alteration in the paternal sperm DNA methylation and the autistic symptoms was found [19]. Consequently, it is highly possible that age-related alteration in DNA methylation of the sperm puts children at risk of developing ASD. While further study should be required, one thing for sure is that sperm aging is not just a matter of fertility.

4 Verification from Animal Models

Effects of paternal aging on the next generation have been studied using mice as the pathological model of neurodevelopmental diseases. Foldi et al. used C57BL/6 J mice by mating 9 male mice at 4 months or 11 male mice at 12–18 months with 4-month-old female mice and conducted behavioral analyses on the offspring (i.e., male and female mice from the young or old father, about 20 F1 in each group) [20]. As a result, there was a slight difference in the behavior of the offspring due to the advanced paternal age. This was found especially in the female offspring with increased anxiety-related behavior, increased exploration, and decreased learning.

In another study, Swiss Albino mice were analyzed up to the grandchild generation [21]. Four-month-old female mice were mated with 10 male mice at 4-month-old or 9 male mice at 15-month-old, and behavior of 15 offspring (F1 mice) from each group was examined; the offspring from old fathers showed abnormal vocal communication by maternal separation and impaired sociability with increased stereotypy and anxiety. When 4-month-old male F1 mice were bred with 4-month-old female mice to obtain F2 offspring (grandchildren), the anxiety behavior observed in F1 mice was restored in F2 mice, but the other abnormal behaviors were still observed. This suggests that some effects of the advanced age of the grandfather persist; some get canceled out in the grandchildren's generation.

Our laboratory has also been conducting similar experimental paradigms using C57BL/6 J mice and found that the offspring of advanced age fathers has abnormality in vocal communication, sensorimotor gating, and spatial learning (our unpublished results). For the vocal communication, we analyzed the maternal separation-induced ultrasonic vocalization (USV) at about 1 week after birth, and the number of calls from the offspring of old male mice was decreased. Our data seem to be different from the results of the abovementioned study using Swiss Albino mice [21]; this could be due to difference in the strain of the mice used. Because even the phenotype of the autism-related gene knockout mice display both increased and decreased level of USV when compared to wild-type mice [22], aberration from typical level of USV in the strain that was used in each study should be considered as vocal communication defect, a phenotype corresponding to human vocal communication.

Epigenetic information, not only at the level of DNA methylation but also of chromatin structures and noncoding RNAs, can be altered in sperm cells. Therefore, the mechanism of how these alterations in epigenetic information can affect the phenotype of future generations is of great interest. Recently, Gingrich's group performed genome-wide methylation profiling of sperm DNA to examine the effect of aging [23]. By comparing DNA methylation patterns obtained from young and old sperm DNA, the authors found hypomethylation in old sperm DNA, especially near the transcription start site (TSS) regions, although the level of DNA methylation was not changed in proximal promoter regions (± 1 kb of TSS). These regions corresponded to CpG island shores, which are known to be associated with the regulation of gene expression [24]. Furthermore, global patterns of DNA methylation in the offspring brain derived from old father were similar to that of old sperm. This suggests the possibility that alterations of sperm DNA methylation established by advanced aging can be retained and may have impact on offspring gene expression. It is generally known that DNA methylation of the sperm genome is erased through reprogramming after fertilization [25], whereas a recent study shows that some genomic regions and imprinted genes elude reprogramming [26]. To examine whether behavioral abnormalities observed in offspring of old father are attributed to alterations of sperm DNA methylation, it will be necessary to investigate the precise mechanism of how these alterations of sperm DNA methylation elude reprogramming process.

Regulation of gene expression during development is also governed by histone modification. Nucleosomal retention of histone proteins such as H3K4me2 and H3K27me3 is reported in sperm genome, especially at loci that are related to regulation of developmental genes [27, 28]. H3K9me2 is an "epigenetic mark" that protects DNA methylation state of the maternal genome from reprograming process after fertilization [29]. It is thus suggested that paternally inherited histone modifications through sperm could contribute to the developmental program of offspring, although a recent study pointed out the situation is not so simple.

A recent study has reported abnormal developmental phenotypes in offspring of transgenic (TG) mice that overexpress histone demethylase *KDM1A* gene, encoding an enzyme demethylating histone H3K4me2 [30]. In these TG mice, human *KDM1A* gene was specifically expressed in the adult testis by using human EF-1α promoter, and basic spermatogenesis was not affected at the level of sperm cell count and apoptotic cell count. Male TG mice were used as founder mice (TG-F1) to generate offspring mice (TG and non-TG F2), a high percentage of which displayed dysplasia in the skin, limb, and craniofacial region with a lower postnatal survival rate, regardless of whether the offspring possessed the paternal transgene. It is important to note that even the grand-offspring (non-TG F3) derived from non-TG F2 male have a high rate of the same abnormal morphogenesis as the F2 generation. These results suggest that the effect of epigenetic dysregulation during spermatogenesis may confer risk factors on sperm, which lasts over the course of two generations and affects the development of offspring.

In the abovementioned study, chromatin immunoprecipitation sequencing analysis of TG F3 sperm revealed altered H3K4me2 profile at CpG-rich sequences within TSS. Contrary to expectation, however, there was less similarity between H3K4me2 profiles in non-TG F3 sperm and TG F3 sperm, suggesting that alteration of sperm H3K4me2 profile does not directly mediate developmental defects across generations. Considering that RNA profile of sperm was similar between TG F3 and non-TG F3, abnormal histone methylation that governs transcription during spermatogenesis in *KDM1A* TG mice might affect sperm transcriptome, possibly at the spermatid stage when the cytoplasm is shared. Therefore, it is reasonable to assume that abnormal RNA profile in sperm can cause developmental anomalies in offspring by altering the gene expression program of initial stages after fertilization. In a previous study on the effect of paternal aging on reproductive success, the authors found, in aged sperm of outbred CF1 mice, decreased mRNA levels of some genes that have key roles in spermatogenesis, sperm motility, and acrosome reaction [31]. Thus, it is reasonable that paternal aging resulted in the poor fertility rate. More interestingly, the authors also reported that fetuses derived from aged male mice (more than 12 months old) showed smaller size and dramatically lower body weight [31]. This finding suggests that sperm from aged male may have abnormal RNA profile that can affect not only male fertility but also fetal development. Since similar correlation between paternal aging and lower birth weight has been reported in human studies [32, 33], the effect of paternal aging should be paid more attention.

For now there is no evidence showing alterations in histone modification during spermatogenesis with advancing age. However, abovementioned pioneering research suggests a possibility that changes in the histone modification-dependent chromatin state during spermatogenesis also lead to changes of RNA profile in sperm, thereby affecting fetal development. Therefore, we can logically assume that the brain function may also be impaired in subsequent generations. Further study is required on the epigenetic changes in sperm that can affect traits of future generations.

5 Conclusion

As introduced above, we now know that paternal aging can have a great influence on the development of future generations. DOHaD hypothesis tended to focus most likely on effects originating from the maternal side, but it is necessary to study in greater detail the effect from the paternal side through the sperm epigenetics. Other than the advanced age, factors such as malnourishment or imbalanced nutrition, smoking, drug exposure, and radiation exposure can cause epigenetic alteration in spermatogenesis (e.g., a review by Anderson et al. [34]). It is our hope that further studies on sperm epigenetics will lead to a more comprehensive understanding of DOHaD theory.

Acknowledgement This work was supported by Grant-in-Aid from MEXT (16H06530).

References

1. Ravelli GP, Stein ZA, Susser MW. Obesity in young men after famine exposure in utero and early infancy. N Engl J Med. 1976;295(7):349–53.
2. Ravelli AC, van der Meulen JH, Michels RP, Osmond C, Barker DJ, Hales CN, et al. Glucose tolerance in adults after prenatal exposure to famine. Lancet. 1998;351(9097):173–7.
3. Roseboom TJ, van der Meulen JH, Osmond C, Barker DJ, Ravelli AC, Schroeder-Tanka JM, et al. Coronary heart disease after prenatal exposure to the Dutch famine, 1944–45. Heart. 2000;84(6):595–8.
4. Barker DJ, Osmond C. Infant mortality, childhood nutrition, and ischaemic heart disease in England and Wales. Lancet. 1986;1(8489):1077–81.
5. Susser ES, Lin SP. Schizophrenia after prenatal exposure to the Dutch hunger winter of 1944–1945. Arch Gen Psychiatry. 1992;49(12):983–8.
6. St Clair D, Xu M, Wang P, Yu Y, Fang Y, Zhang F, et al. Rates of adult schizophrenia following prenatal exposure to the Chinese famine of 1959-1961. JAMA. 2005;294(5):557–62.
7. Statistics and Information Department, Ministry of Health, Labor and Welfare. Tokyo: Vital Statistics of Japan. Available from: http://www.mhlw.go.jp/english/database/db-hw/vs01.html
8. Weintraub K. The prevalence puzzle: autism counts. Nature. 2011;479(7371):22–4.
9. Reichenberg A, Gross R, Weiser M, Bresnahan M, Silverman J, Harlap S, et al. Advancing paternal age and autism. Arch Gen Psychiatry. 2006;63(9):1026–32.
10. Sandin S, Schendel D, Magnusson P, Hultman C, Suren P, Susser E, et al. Autism risk associated with parental age and with increasing difference in age between the parents. Mol Psychiatry. 2015;21:693–700.
11. Zammit S, Allebeck P, Dalman C, Lundberg I, Hemmingson T, Owen MJ, et al. Paternal age and risk for schizophrenia. Brit J Psychiat. 2003;183:405–8.
12. Malaspina D, Harlap S, Fennig S, Heiman D, Nahon D, Feldman D, et al. Advancing paternal age and the risk of schizophrenia. Arch Gen Psychiatry. 2001;58(4):361–7.
13. Sanders SJ, Murtha MT, Gupta AR, Murdoch JD, Raubeson MJ, Willsey AJ, et al. De novo mutations revealed by whole-exome sequencing are strongly associated with autism. Nature. 2012;485(7397):237–41.
14. Neale BM, Kou Y, Liu L, Ma'ayan A, Samocha KE, Sabo A, et al. Patterns and rates of exonic de novo mutations in autism spectrum disorders. Nature. 2012;485(7397):242–5.
15. O'Roak BJ, Vives L, Girirajan S, Karakoc E, Krumm N, Coe BP, et al. Sporadic autism exomes reveal a highly interconnected protein network of de novo mutations. Nature. 2012;485(7397):246–50.
16. Kong A, Frigge ML, Masson G, Besenbacher S, Sulem P, Magnusson G, et al. Rate of de novo mutations and the importance of father's age to disease risk. Nature. 2012;488(7412):471–5.
17. Jenkins TG, Aston KI, Pflueger C, Cairns BR, Carrell DT. Age-associated sperm DNA methylation alterations: possible implications in offspring disease susceptibility. PLoS Genet. 2014;10(7):e1004458.
18. Lee HY, Jung SE, Oh YN, Choi A, Yang WI, Shin KJ. Epigenetic age signatures in the forensically relevant body fluid of semen: a preliminary study. Forensic Sci Int Genet. 2015;19:28–34.
19. Feinberg JI, Bakulski KM, Jaffe AE, Tryggvadottir R, Brown SC, Goldman LR, et al. Paternal sperm DNA methylation associated with early signs of autism risk in an autism-enriched cohort. Int J Epidemiol. 2015;44(4):1199–210.

20. Foldi CJ, Eyles DW, McGrath JJ, Burne TH. Advanced paternal age is associated with alterations in discrete behavioural domains and cortical neuroanatomy of C57BL/6J mice. Eur J Neurosci. 2010;31(3):556–64.
21. Sampino S, Juszczak GR, Zacchini F, Swiergiel AH, Modlinski JA, Loi P, et al. Grand-paternal age and the development of autism-like symptoms in mice progeny. Transl Psychiatry. 2014;4:e386.
22. de la Torre-Ubieta L, Won HJ, Stein JL, Geschwind DH. Advancing the understanding of autism disease mechanisms through genetics. Nat Med. 2016;22(4):345–61.
23. Milekic MH, Xin Y, O'Donnell A, Kumar KK, Bradley-Moore M, Malaspina D, et al. Age-related sperm DNA methylation changes are transmitted to offspring and associated with abnormal behavior and dysregulated gene expression. Mol Psychiatry. 2015;20(8):995–1001.
24. Irizarry RA, Ladd-Acosta C, Wen B, Wu Z, Montano C, Onyango P, et al. The human colon cancer methylome shows similar hypo- and hypermethylation at conserved tissue-specific CpG island shores. Nat Genet. 2009;41(2):178–86.
25. Reik W, Dean W, Walter J. Epigenetic reprogramming in mammalian development. Science. 2001;293(5532):1089–93.
26. Kobayashi H, Sakurai T, Imai M, Takahashi N, Fukuda A, Yayoi O, et al. Contribution of intragenic DNA methylation in mouse gametic DNA methylomes to establish oocyte-specific heritable marks. PLoS Genet. 2012;8(1):e1002440.
27. Hammoud SS, Nix DA, Zhang H, Purwar J, Carrell DT, Cairns BR. Distinctive chromatin in human sperm packages genes for embryo development. Nature. 2009;460(7254):473–8.
28. Brykczynska U, Hisano M, Erkek S, Ramos L, Oakeley EJ, Roloff TC, et al. Repressive and active histone methylation mark distinct promoters in human and mouse spermatozoa. Nat Struct Mol Biol. 2010;17(6):679–87.
29. Nakamura T, Liu YJ, Nakashima H, Umehara H, Inoue K, Matoba S, et al. PGC7 binds histone H3K9me2 to protect against conversion of 5mC to 5hmC in early embryos. Nature. 2012;486(7403):415–9.
30. Siklenka K, Erkek S, Godmann M, Lambrot R, McGraw S, Lafleur C, et al. Disruption of histone methylation in developing sperm impairs offspring health transgenerationally. Science. 2015;350(6261):aab2006.
31. Katz-Jaffe MG, Parks J, McCallie B, Schoolcraft WB. Aging sperm negatively impacts in vivo and in vitro reproduction: a longitudinal murine study. Fertil Steril. 2013;100(1):262–8 e1–2.
32. Alio AP, Salihu HM, McIntosh C, August EM, Weldeselasse H, Sanchez E, et al. The effect of paternal age on fetal birth outcomes. Am J Mens Health. 2012;6(5):427–35.
33. Reichman NE, Teitler JO. Paternal age as a risk factor for low birthweight. Am J Public Health. 2006;96(5):862–6.
34. Anderson D, Schmid TE, Baumgartner A. Male-mediated developmental toxicity. Asian J Androl. 2014;16(1):81–8.

Part IV

Clinical Significance

Preemptive Epigenetic Medicine Based on Fetal Programming

Takeo Kubota

Abstract

The developmental origins of health and disease (DOHaD) refers to the concept that environmental stress during pregnancy alters the programmed fetal development and subsequently causes disorders, such as cardiovascular and metabolic diseases, in adulthood. Epigenetics is a gene regulation mechanism that does not depend on DNA sequence but on chemical modifications of DNA. Several lines of evidence suggest that environmental stress in the fetal period alters the epigenetic state of genes, leading to permanent gene dysregulation, which may be associated with disorders that emerge after birth. Such stresses include malnutrition, which may be associated with type 2 diabetes, and mental stress, which may be associated with neurodevelopmental disorders. It has also been demonstrated that environmental stress-induced epigenetic alterations can be transmitted to the next generation via disease phenotypes. However, since epigenetic modification is an internal system based on attachment and detachment of chemical residues on a DNA sequence, it is reversible and potentially treatable. In fact, recent studies demonstrated that some drugs and early interventions are effective at preventing epigenetic disorders. Therefore, preventive and preemptive medicine is possible for disorders caused by alterations in programming during fetal and early periods.

Keywords

DOHaD · Epigenetics · Epigenome · Reversibility · Preemptive medicine

1 Introduction

It is known that environmental stress during pregnancy (e.g., drug exposure or viral infections) can cause congenital anomalies in the fetus. For example, valproic acid, a drug commonly used for epilepsy and psychiatric disorders, is associated with the occurrence of spina bifida, congenital heart defects, and cleft lip/plate. Rubella viral infection is known to increase the risk of congenital heart defects, deafness, and congenital cataract. Similarly, bisphenol A, a plastic plasticizer that is an environmental chemical, is known to increase the risk of urological anomalies, such as cryptorchidism and hypospadias, in fetuses [1]. Because of these findings, pharmaceutical companies disclose fetus-related side effects of prescription drugs in patient package inserts; in addition, the rubella vaccination is now recommended prior to pregnancy in young women (Fig. 1, top).

T. Kubota (✉)
Faculty of Child Studies, Seitoku University, Matsudo, Chiba, Japan
e-mail: kubota.takeo@seitoku.ac.jp

© Springer Nature Singapore Pte Ltd. 2018

T. Kubota, H. Fukuoka (eds.), *Developmental Origins of Health and Disease (DOHaD)*, Advances in Experimental Medicine and Biology 1012, https://doi.org/10.1007/978-981-10-5526-3_9

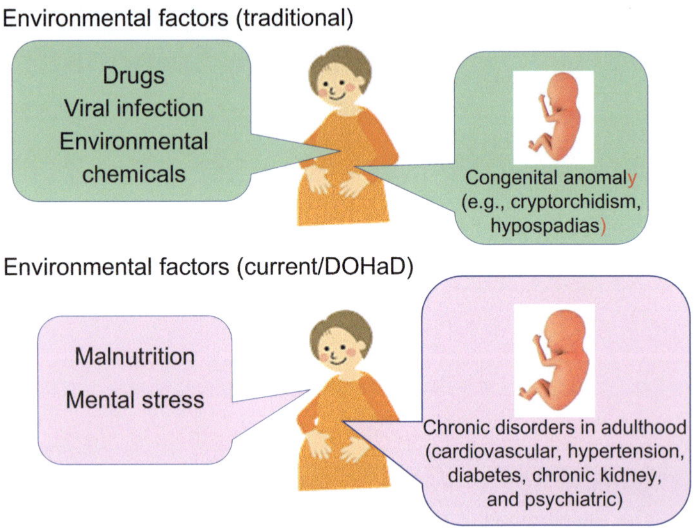

Fig. 1 Environmental factors, such as drug exposure and viral infection during pregnancy, are known to cause congenital anomalies in fetuses. Under the current concept of the developmental origins of health and disease (DOHaD), maternal environmental stress during pregnancy can cause adult-onset disorders in the fetus

It was recently suggested that environmental stress during pregnancy not only causes congenital anomalies that are immediately detectable at birth but also adult disease, such as diabetes and hypertension. This led to the hypothesis that adult diseases could originate during the fetal period. This hypothesis was demonstrated in part; the theory is now referred to as the developmental origins of health and disease (DOHaD) (Fig. 1, bottom). Obstetricians and pediatricians have not attended to DOHaD because the target diseases are not urgent but rather chronic. However, they must understand the significance of DOHaD because if they do not intervene at all, the number of patients with adult diseases will increase, and the economic power of countries will weaken over several decades. Therefore, the Japan Society for the Promotion of Science recently recognized the DOHaD concept and planned to implement a new national project [2].

Based on the trends described above, I introduce the social background behind the proposed DOHaD theory, update the biological understanding of DOHaD, and present a future perspective of medicine and society based on the DOHaD theory in this chapter.

2 Social Background of DOHaD

The miniskirt was introduced to Japan by Europe in the early 1970s. Since then, young Japanese women have tried to keep their legs slim and became interested in intentional dieting. Dieting during early pregnancy, perhaps even before the pregnancy is detected, could potentially cause fetal malnutrition. Dieting may be one cause of the current trend of decreased birth weights in Japan [3]. The United Nations International Children's Emergency Fund (UNICEF) reported that Japan is the second highest ranking country for low birth weight (live babies born weighing less than 2500 g) among developed countries, only after Greece [4].

Recent epidemiological studies reported that, in the Netherlands and China in the 1940s–1960s, offspring born to mothers exposed to famine during their first and second trimesters had lower birth weights compared with offspring born to mothers not exposed to famine; these underweight offspring had increased risks of cardiovascular, metabolic, and psychiatric disorders [5–7].

Furthermore, another epidemiological study suggested that malnutrition during the fetal period increased the risk of chronic kidney disease, namely, microalbuminuria; this may be due to abnormalities during the period of midgestation in which nephron number increases, leading to glomerular hypertension and hyperfiltration [8, 9]. In fact, renal biopsies performed on individuals who were born with extremely low birth weight revealed compensatory glomerular hyperfiltration with proteinuria and subsequent hypertension [10].

Therefore, the current trend of fetal exposure to hyponutrition, which may lead to low birth weight, may increase the future number of adult patients with chronic disorders in Japan. This expectation is not limited to Japan but can also be applied to many developing countries in Asia, the Middle East, and Africa.

3 Biological Basis of DOHaD

3.1 Epigenetic Mechanisms

In mammals, the expression of genes essential for development during embryogenesis [11] as well as differentiation of cells, including neural cells [12, 13], is controlled in part by epigenetics. Therefore, understanding epigenetics, including DNA methylation, histone modifications, and microRNA, is important for elucidating pathogenic pathways in DOHaD-related disorders [14–19]. In fact, there are a number of congenital neurodevelopmental disorders (NDs) caused by epigenetic failures [18–29].

DNA methylation is a step in epigenetic gene control, which occurs by the DNA methyltransferase (DNMT)-mediated addition of a methyl group (CH_3) to CpG dinucleotides. Among the DNMTs, mutations in DNMT3B are known to cause immunodeficiency-centromeric instability-facial anomalies (ICF) syndrome [20–22]. This indicates that DNMT is essential for immunological and neurological development.

Methyl-CpG-binding domain (MBD) proteins are also important for epigenetic regulation. Mutations in the methyl-CpG-binding protein 2 (*MECP2*) gene is known to cause Rett syndrome (RTT), which is characterized by seizures, ataxic gait, and autistic features [23, 24]. These characteristics of RTT are reproducibly displayed in *Mecp2* knockout mice [25, 26] and patient-derived mutation knock-in mice [27, 28]. Since these features are also found in neuronal cell-specific *Mecp2* knockout mice [25, 29], abnormalities in neurons are thought to be a main cause for RTT neurological features. In fact, misregulation of neuronal genes, which encode brain-derived neurotrophic factor, distal-less homeobox 5, inhibitors of DNA binding, a member of the corticotropin-releasing factor family, insulin-like growth factor-binding protein-3, cyclin-dependent kinase-like 1, protocadherin beta 1, and protocadherin 7, due to lack of MeCP2 is found in neurons [30–34], indicating that proper epigenetic gene regulation in neurons is essential for normal brain development.

3.2 Mechanisms of Mental Stress-Induced Epigenetic Changes

Both environmental factors (e.g., toxins, infections) and genetic factors (e.g., mutations in synaptic molecules) are thought to be involved in NDs [35–37]. However, a biological link between environmental and genetic factors has not been identified. Epigenetics may be the bridge between these factors, thereby contributing to disease development [14]. In addition to the congenital epigenetic defects described above, environmental factors, such as malnutrition [29, 38, 39], drug exposure [40–44], mental stress during the neonatal period [45], and neuronal stimulation [46], are also thought to alter the epigenetic status and thereby affect brain function. Therefore, it is intriguing to speculate that acquired NDs may be the result of epigenetic dysregulation in either over-suppression or under-suppression of genes caused by environmental factors (Fig. 2).

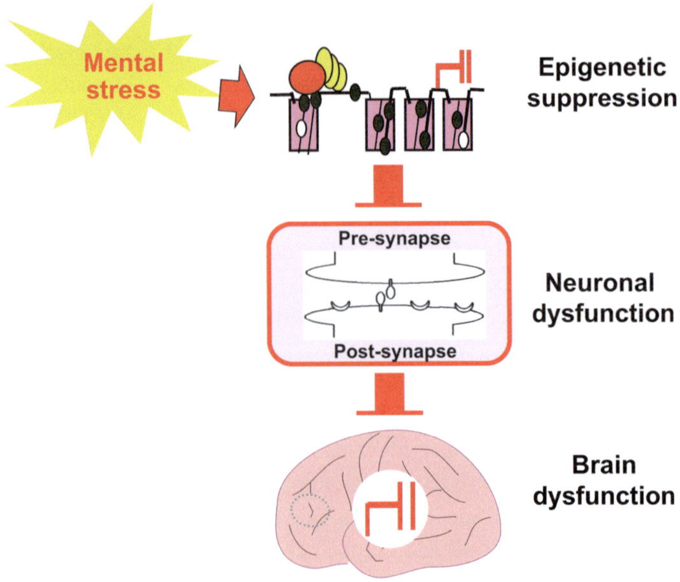

Fig. 2 Mutations in genes encoding proteins associated with neuronal function are found in a subset of patients with neurodevelopmental disorders (NDs). Mental stress affects the function of epigenetic regulation-associated genes. All of these genetic alterations and epigenetic misregulation by mental stress lead to altered brain function

A sensational report suggested that short-term mental stress after birth alters the epigenetic status in the brain and results in persistent abnormal behavior [47]. Specifically, low levels of maternal care within the first week of life increased DNA methylation at the promoter of the glucocorticoid receptor (*GR*) gene, which is also known as the nuclear receptor subfamily 3 group C member 1 (*NR3C1*) gene, in the hippocampus of rat pups, which subsequently suppressed *NR3C1* expression. In contrast, promoter methylation decreased in the hippocampus of the offspring of mothers who engaged in high maternal care during the same period [48]. This paradigm provided an animal model. Postmortem analysis of the hippocampus of suicide victims with a history of childhood abuse revealed the presence of hypermethylation of the *NR3C1* promoter in combination with subsequent decreased *NR3C1* expression [47]. This finding suggests that the adverse effects of early-life stress on DNA methylation may last throughout life [48] and also indicates that neurodevelopmental problems due to childhood neglect and maltreatment may arise from epigenetic dysregulation caused by environmental stress in early life.

Epigenetic mechanisms may also play a role in the pathophysiology of post-traumatic stress disorder (PTSD) [49, 50]. In one recent study, increased DNA methylation within Alu repetitive elements was found in serum DNA (cell-free DNA) of PTSD cases in US military service members who were deployed in Iraq and Afghanistan [51]. Alu hypermethylation can be interpreted to be a response to psychological stress [51]. On the contrary, increased DNA methylation was found in long interspersed nuclear element-1 (LINE-1 or L1) repetitive elements in non-PTSD cases after deployment. These findings suggest that psychological stress incurred during deployment may induce DNA methylation in a subset of repetitive sequences in the human genome, which may protect against PTSD [51].

It has also been reported that DNA methylation differences are greater in normal older monozygotic twins than in normal younger twins [52]. DNA methylation differences are also found in monozygotic twins with discordant RTT severity, with subsequent differences in neuronal gene expression [53]. These reports indicate that aging or environmental factors may affect epigenomic patterns and psychological status in humans.

3.3 Mechanisms of Malnutrition-Induced Epigenetic Changes

As was briefly mentioned in the Social Background of DOHaD section, in Japan, dieting is now popular among young women, both before and during pregnancy. This is partly because obstetric physicians recommend dieting to minimize pregnancy weight gain in order to reduce the risk of gestational diabetes and the risk of delivery failure. This may contribute to the decline in birth weights over the past 20 years in Japan [3]. This generation of individuals with lower birth weight is expected to have an increased risk of metabolic disorders (e.g., obesity, type 2 diabetes) and psychiatric disorders based on data from current epidemiological studies of populations affected by famine in the Netherlands and China [6, 7] per the DOHaD [54].

Animal studies have demonstrated that malnutrition during the fetal period causes a hypomethylation imprint on the peroxisome proliferator-activated receptor alpha (*PPARα*) gene in rat liver [55]. Similar DNA methylation changes have been identified in peripheral blood samples from people who suffered malnutrition during a period of famine in the Netherlands [56]. It has also been reported that assisted reproductive technologies (e.g., in vitro fertilization and intracytoplasmic sperm injection), which are now widely used due to increased age at the time of marriage, lead to decreased DNA methylation status at multiple maternally methylated loci in the human genome [57, 58]. These findings suggest that environmental stress during gamatogenetic, embryogenetic and fatal periods can alter the epigenetic state of fertilized eggs, embryos, and fetuses.

4 Transgenerational Epigenetic Inheritance

4.1 Mechanisms of Transgenerational Epigenetic Inheritance

In accordance with the current understanding of biological inheritance, it has been thought that one's acquired characteristics should not be transmitted to the next generation (i.e., children). For example, based on this notion of Darwinian inheritance, acquired changes induced by harmful habits (e.g., imbalanced diet) during the father's lifetime should not be transmitted to his children. However, it has recently been reported that such undesirable acquired traits (e.g., imbalanced diet) are transmitted to the next generation through DNA methylation changes in the father's sperm [59].

It is known that epigenetic marks allow the mitotic transmission of gene activity states from one cell to its daughter cells. A fundamental question in epigenetics is whether these marks can also be transmitted meiotically through the germline. In mammals, epigenetic marks should be cleared by demethylating factors, such as cytidine deaminases (e.g., AID, APOBEC1), and reestablished in each generation. However, this clearing is incomplete at some loci in the genome of several model organisms, possibly due to a deficiency in demethylating factors (e.g., AID) [60]. Therefore, based on this phenomenon, transgenerational epigenetic inheritance, which refers to the germline transmission of an environment-induced epigenetic mark [61, 62], may provide direct biological evidence for Lamarckism (i.e., the hypothesis that an organism can pass on characteristics acquired during its lifetime to its offspring or the hypothesis of the heritability of acquired characteristics).

4.2 Heritable Germline Epimutation

It is important to draw a distinction between transgenerational epigenetic inheritance and heritable germline epimutation. Transgenerational epigenetic inheritance is independent of the DNA sequence, whereas heritable germline epimutation is a direct consequence of a *cis*-acting epigenetic alteration due to a specific DNA sequence (e.g., unstable CGG repeat within exon 1 of *FMR1*) [62]. Such specific-sequence-driven epimutation (i.e., hypermethylation) is observed in an affected mother and her son in families with fragile X syndrome [63]. This inheritance pattern is also observed in other species, such as *Caenorhabditis elegans*. Therefore, the heritability does not necessarily involve transgenerational epigenetic inheritance, as the methylated state is once cleared on passage through the germline and is then reestablished according to a certain sequence during zygotic genome activation [63].

4.3 Significance of Transgenerational Epigenetic Inheritance in Plants

When we think of the significance of transgenerational epigenetic inheritance, one can imagine that it would be more beneficial for plants (sessile organisms) than animals (movable organisms) to pass on the information that they acquire about their environment to their progeny, as seeds are often dispersed locally [64]. This may be due to the fact that, compared with plants, animals can more readily move from place to place in response to environmental changes (i.e., climate). Therefore, this inheritance system may be less essential in animals than in plants. Indeed, plants have an extensive repertoire of epigenetic regulation involving mechanisms based on DNA methylation, histone modification, and RNA. DNA methylation and chromatin modifications are guided in a sequence-specific manner by small RNAs in *Arabidopsis* [65], and such small RNAs can be regulated by environmental stress [66].

Even when isogenic plants are grown under uniformly benign conditions, differences in DNA methylation have been shown to accumulate in a set of *Arabidopsis thaliana* mutation accumulation lines [67, 68]; similar findings have been observed in human identical twins [52, 53]. Furthermore, whole-genome methylome sequencing of individual plants in Generations 3 and 31 has revealed that polymorphic methylation at individual cytosine sites is several orders of magnitude more frequent than DNA mutation [69], indicating that spontaneous or induced variation in the efficiency of the methylation machinery could underlie heritable epigenetic changes associated with phenotypic differences that could ultimately become subject to Darwinian selection [70]. As mentioned above, this phenomenon is also observed in humans, which may be the genetic mechanism for variable expressivity, a type of phenotypic variability between family members who share the same disease-causing mutation, which may be based on stochastic changes in epigenetic modifications [69].

4.4 Transgenerational Epigenetic Inheritance in Mammals

In mammals, transgenerational inheritance of epigenetic marks was first demonstrated using a specific mouse strain with the murine agouti-viable yellow (A(vy)) allele. The methylation status at the *Axin (Fu)* locus in mature sperm reflects the methylation state of the allele in the somatic tissue of the animal; the methylation status is linked to the shape of the animal's tail. Therefore, the methylation status does not undergo epigenetic reprogramming during gametogenesis [71].

Environmental chemicals and nutritional challenges have been associated with transgenerational epigenetic inheritance in animal models. However, it is often difficult to dissect evidence of transmission of epigenetic marks per se from transmission of the exposure itself [72, 73]. Therefore, transgenerational effects should be distinguished from parental and grandparental effects. In addition to contributing their DNA, parents can influence their offspring in many other ways, for example, by contributing bioactive molecules in the egg and sperm cytoplasm and by providing nutrients and hormonal information during embryogenesis. Malnutrition during pregnancy not only affects the pregnant mother and fetus but also the fetus' primordial germ cells, which can lead to phenotypic changes in the grandchildren (second generation). Methylation status of the *Axin (Fu)* locus is linked to hair color of the mice and can be altered by dietary folic acid supplementation; this status was inherited over two generations but was lost by the third generation [74]. These findings indicate that the acquired epigenetic information is not inherited transgenerationally because a specific diet may lead to parental and grandparental effects or that the *Axin (Fu)* locus is resistant to environmentally induced acquisition of new germline epigenetic information [74].

However, a further study revealed that transgenerational effects of environmental toxins (e.g., the antiandrogenic compound vinclozolin, a commonly used fungicide; the estrogenic compound methyoxychlor) have been demonstrated in the fourth generation (F4) of rats with decreased spermatogenic capacity and increased male infertility; the effects on reproduction correlate with altered DNA methylation patterns in the germline [75]. It has also been demonstrated that plastic-derived endocrine disrupters (e.g., bisphenol A) increase the risk of pubertal abnormalities in the F3 generation in rats and alter DNA methylation in the F3 generation sperm [76]. Therefore, transgenerational epigenetic inheritance up to the third generation is suggested. Transmission of environment (heat shock)-induced epigenetic change (defective chromatin state) by the next generation has also been confirmed in *Drosophila* [77].

As described above, short-term mental stress due to maternal-neonate separation immediately after birth alters the epigenetic status in the brain of the neonate and results in persistent abnormal behavior [45]. It has further been demonstrated that such environment-induced epigenetic changes occur not only in the brain but also in the sperm; thus, aberrant environment-induced epigenetic marks acquired in one generation can be inherited by the next generation [78]. In other words, chronic maternal separation altered behaviors, and DNA methylation in the promoter of several genes in the germline of maternally separated mice and the epigenetic changes were then observed in the brains of the offspring with altered gene expression, which included decreased expression of corticotropin-releasing factor receptor 2 (*Crfr2*) in the amygdala and hypothalamus [78]. In this study, abnormal behavior was observed even in the third generation, and altered DNA methylation in the CpG islands of *Mecp2*, cannabinoid receptor-1 (*Cb1*), and *Crfr2* was observed in F1 sperm and F2 brain [78].

In a separate study, chronic maternal separation increased cytosine methylation of the estrogen receptor (*Er*)-alpha1b gene promoter, indicating that individual differences in maternal behavior are epigenetically transmitted from the mother to her female offspring [79]. Furthermore, third generation male rats exposed to vinclozolin respond differently to mental stress during adolescence; they also showed altered gene expression in the cortex and CA1 regions of the brain [80]. These findings provide biological evidence suggesting that environmental factors, including traumatic experiences in early life, are risk factors for the development of behavioral and emotional disorders not only in one generation but also in successive generations.

5 Perspectives

As mentioned above, mental stress in the first week of life causes epigenetic abnormalities in the brains of mice. Conversely, several mouse studies have demonstrated that brain-stimulating conditions can ameliorate behavioral abnormalities. Such studies have been performed using a mouse model of RTT. For example, environmental enrichment (EE), consisting of larger-sized home cages with a variety of objects, including running wheels, led to improved motor coordination and decreased anxiety-related behavior in heterozygous *Mecp2*$^{+/-}$

female mice [81, 82]. EE also improved locomotor activity with reduced ventricular volume and restored the expression of synaptic markers, such as synaptophysin and postsynaptic density protein 95 in the hypothalamus and syntaxin 1a and synaptotagmin in the cortex of hemizygous *Mecp2*$^{-/y}$ male mice [83, 84].

It is difficult to cure patients with congenital psychiatric disorders caused by mutations that encode neuronal molecules, as it is difficult to distribute gene products to the appropriate brain regions and at the appropriate time during brain development. However, it was recently demonstrated that the epigenetic disorder RTT may be an exception, partly because *MeCP2* does not encode the molecule that functions as "one of the parts of the brain," but encodes a "lubricant" that is distributed to many places (genes) in the genome. MeCP2 functions as a transcriptional repressor and acts as a "noise reducer" in neuronal cells; these functions contribute to maturation of the brain and are active at a relatively later period during brain development. This is supported by recent findings that MeCP2 targets a relatively large number of synaptic molecules [29–33] and that postnatal loss of MeCP2 in the forebrain recapitulates features of RTT in mice [28]. Indeed, reintroduction of *MECP2* into *Mecp2* null mice after birth, as well as before birth, is sufficient to rescue various RTT symptoms [85, 86]. Furthermore, restoration of MeCP2 function in astrocytes (not neurons) substantially improves locomotion, anxiety levels, and respiratory abnormalities in hemizygous *Mecp*$^{-/Y}$ male mice, along with restoring dendritic morphology [87]. These results suggest that upregulation of *MECP2*, possibly mediated by drug treatment, might help to improve brain function in patients with RTT and further indicate that NDs other than RTT caused by epigenetic abnormalities may also be treatable.

Epigenetic markings provide a "memory of past experiences." These markings persist during the life span of an individual and are sometimes transmitted to the offspring. We still do not know whether epigenetics, the genomic reaction to environmental factors, is beneficial or harmful; it likely acts in both directions [88]. If it is the latter, reversing or preventing the harmful epigenetic signature is important. This will be the basis of preventive and preemptive medicine [89].

References

1. Fernández MF, Arrebola JP, Jiménez-Díaz I, Sáenz JM, Molina-Molina JM, Ballesteros O, et al. Bisphenol A and other phenols in human placenta from children with cryptorchidism or hypospadias. Reprod Toxicol. 2016;59:89. https://doi.org/10.1016/j.reprotox.2015.11.002.
2. Workshop report for future scientific technology "Elaboration of preemptive medicine from the fetal and neonatal point of view" by Center for Research and Development Strategy (in Japanese). https://www.jst.go.jp/crds/pdf/2013/WR/CRDS-FY2013-WR-14.pdf. Checked on Feb 20th 2017.
3. Gluckman PD, Seng CY, Fukuoka H, Beedle AS, Hanson MA. Low birthweight and subsequent obesity in Japan. Lancet. 2007;369(9567):1081. https://doi.org/10.1016/S0140-6736(07)60524-8.
4. Countries ranked by Low-birthweight babies (% of births) – IndexMundi. http://www.indexmundi.com Indicators›Health›Nutrition. Checked on Feb 20th 2017.
5. Lumey LH. Decreased birthweights in infants after maternal in utero exposure to the Dutch famine of 1944–1945. Paediatr Perinat Epidemiol. 1992;6:240–53.
6. Painter RC, de Rooij SR, Bossuyt PM, Simmers TA, Osmond C, Barker DJ, et al. Early onset of coronary artery disease after prenatal exposure to the Dutch famine. Am J Clin Nutr. 2006;84(2):322–7.
7. St Clair D, Xu M, Wang P, Yu Y, Fang Y, Zhang F, et al. Rates of adult schizophrenia following prenatal exposure to the Chinese famine of 1959-1961. JAMA. 2005;294(5):557–62.
8. Painter RC, Roseboom TJ, van Montfrans GA, Bossuyt PM, Krediet RT, Osmond C, et al. Microalbuminuria in adults after prenatal exposure to the Dutch famine. J Am Soc Nephrol. 2005;16(1):189.
9. Hayashi A, Santo Y, Satomura K. Proteinuria and glomerular hypertrophy in extremely low-birthweight children. Pediatr Int. 2014;56(6):860–84. https://doi.org/10.1111/ped.12466.
10. Brenner BM, Lawler EV, Mackenzie HS. The hyperfiltration theory: a paradigm shift in nephrology. Kidney Int. 1996;49(6):1774.
11. Li E, Beard C, Jaenisch R. Role for DNA methylation in genomic imprinting. Nature. 1993;366(6453):362–5.
12. Takizawa T, Nakashima K, Namihira M, Ochiai W, Uemura A, Yanagisawa M, et al. DNA methylation is a critical cell-intrinsic determinant of astrocyte differentiation in the fetal brain. Dev Cell. 2001;1(6):749–58.

13. Sakashita K, Koike K, Kinoshita T, Shiohara M, Kamijo T, Taniguchi S, et al. Dynamic DNA methylation change in the CpG island region of p15 during human myeloid development. J Clin Invest. 2001;108(8):1195–204.
14. Qiu J. Epigenetics: unfinished symphony. Nature. 2006;441(7090):143–5.
15. Abel T, Zukin RS. Epigenetic targets of HDAC inhibition in neurodegenerative and psychiatric disorders. Curr Opin Pharmacol. 2008;8(1):57–64. https://doi.org/10.1016/j.coph.2007.12.002.
16. Urdinguio RG, Sanchez-Mut JV, Esteller M. Epigenetic mechanisms in neurological diseases: genes syndromes and therapies. Lancet Neurol. 2009;8(11):1056–72. https://doi.org/10.1016/S1474-4422(09)70262-5.
17. Wu H, Tao J, Chen PJ, Shahab A, Ge W, Hart RP, et al. Genome-wide analysis reveals methyl-CpG-binding protein 2-dependent regulation of microRNAs in a mouse model of Rett syndrome. Proc Natl Acad Sci U S A. 2010;107:18161–6. https://doi.org/10.1073/pnas.1005595107.
18. Kubota T, Das S, Christian SL, Baylin SB, Herman JG, Ledbetter DH. Methylation-specific PCR simplifies imprinting analysis. Nat Genet. 1997;16:16–7.
19. Kubota T, Wakui K, Nakamura T, Ohashi H, Watanabe Y, Yoshino M, , et al. Proportion of the cells with functional X disomy is associated with the severity of mental retardation in mosaic ring X turner syndrome females. Cytogenet Genome Res 2002;99(1–4):276–284.
20. Okano M, Bell DW, Haber DA, Li E. DNA methyltransferases Dnmt3a and Dnmt3b are essential for de novo methylation and mammalian development. Cell. 1999;99(3):247–57.
21. Shirohzu H, Kubota T, Kumazawa A, Sado T, Chijiwa T, Inagaki K, et al. Three novel DNMT3B mutations in Japanese patients with ICF syndrome. Am J Med Genet. 2002;112(1):31–7.
22. Kubota T, Furuumi H, Kamoda T, Iwasaki N, Tobita N, Fujiwara N, , et al. ICF syndrome in a girl with DNA hypomethylation but without detectable DNMT3B mutation. Am J Med Genet A .2004;129A(3):290–293.
23. Amir RE, Van den Veyver IB, Wan M, Tran CQ, Francke U, Zoghbi HY. Rett syndrome is caused by mutations in X-linked MECP2 encoding methyl-CpG-binding protein 2. Nat Genet. 1999;23(2):185–8.
24. Chunshu Y, Endoh K, Soutome M, Kawamura R, Kubota T. A patient with classic Rett syndrome with a novel mutation in MECP2 exon 1. Clin Genet. 2006;70(6):530–1.
25. Guy J, Hendrich B, Holmes M, Martin JE, Bird A. A mouse Mecp2-null mutation causes neurological symptoms that mimic Rett syndrome. Nat Genet. 2001;27(3):322–36.
26. Chen RZ, Akbarian S, Tudor M, Jaenisch R. Deficiency of methyl-CpG binding protein-2 in CNS neurons results in a Rett-like phenotype in mice. Nat Genet. 2001;27(3):327–31.
27. Shahbazian M, Young J, Yuva-Paylor L, Spencer C, Antalffy B, Noebels J, et al. Mice with truncated MeCP2 recapitulate many Rett syndrome features and display hyperacetylation of histone H3. Neuron. 2002;35(2):243–54.
28. Ebert DH, Gabel HW, Robinson ND, Kastan NR, Hu LS, Cohen S, et al. Activity-dependent phosphorylation of MeCP2 threonine 308 regulates interaction with NCoR. Nature. 2013;499(7458):341–5. https://doi.org/10.1038/nature12348.
29. Gemelli T, Berton O, Nelson ED, Perrotti LI, Jaenisch R, Monteggia LM. Postnatal loss of methyl-CpG binding protein 2 in the forebrain is sufficient to mediate behavioral aspects of Rett syndrome in mice. Biol Psychiatry. 2006;59(5):468–76.
30. Chen WG, Chang Q, Lin Y, Meissner A, West AE, Griffith EC, et al. Derepression of BDNF transcription involves calcium-dependent phosphorylation of MeCP2. Science. 2003;302(5646):885–9.
31. Martinowich K, Hattori D, Wu H, Fouse S, He F, Hu Y, et al. DNA methylation-related chromatin remodeling in activity-dependent BDNF gene regulation. Science. 2003;302(5646):890–3.
32. Horike S, Cai S, Miyano M, Cheng JF, Kohwi-Shigematsu T. Loss of silent-chromatin looping and impaired imprinting of DLX5 in Rett syndrome. Nat Genet. 2005;37(1):31–40.
33. Itoh M, Ide S, Takashima S, Kudo S, Nomura Y, Segawa M, et al. Methyl CpG-binding protein 2 (a mutation of which causes Rett syndrome) directly regulates insulin-like growth factor binding protein 3 in mouse and human brains. J Neuropathol Exp Neurol. 2007;66(2):117–23.
34. Miyake K, Hirasawa T, Soutome M, Itoh M, Goto Y, Endoh K, et al. The protocadherins, PCDHB1 and PCDH7, are regulated by MeCP2 in neuronal cells and brain tissues: implication for pathogenesis of Rett syndrome. BMC Neurosci. 2011;12:81. https://doi.org/10.1186/1471-2202-12-81.
35. Fombonne E. Epidemiology of pervasive developmental disorders. Pediatr Res. 2009;65(6):591–8.
36. Persico AM, Bourgeron T. Searching for ways out of the autism maze: genetic, epigenetic and environmental clues. Trends Neurosci. 2006;29(7):349–58.
37. Zafeiriou DI, Ververi A, Vargiami E. Childhood autism and associated comorbidities. Brain Dev. 2007;29(5):257–72.
38. Burdge GC, Lillycrop KA, Phillips ES, Slater-Jefferies JL, Jackson AA, Hanson MA. Folic acid supplementation during the juvenile-pubertal period in rats modifies the phenotype and epigenotype induced by prenatal nutrition. J Nutr. 2009;139(5):1054–60. https://doi.org/10.3945/jn.109.104653.
39. Lillycrop KA, Slater-Jefferies JL, Hanson MA, Godfrey KM, Jackson AA, Burdge GC. Induction of altered epigenetic regulation of the hepatic glucocorticoid receptor in the offspring of rats fed a protein-restricted diet during pregnancy suggests that reduced DNA methyltransferase-1 expression is involved in

impaired DNA methylation and changes in histone modifications. Br J Nutr. 2007;97(6):1064–73.
40. Tsankova NM, Berton O, Renthal W, Kumar A, Neve RL, Nestler EJ. Sustained hippocampal chromatin regulation in a mouse model of depression and antidepressant action. Nat Neurosci. 2006;9(4):519–25.
41. Jessberger S, Nakashima K, Clemenson GD Jr, Mejia E, Mathews E, Ure K, et al. Epigenetic modulation of seizure-induced neurogenesis and cognitive decline. J Neurosci. 2007;27(22):5967–75.
42. Dong E, Nelson M, Grayson DR, Costa E, Guidotti A. Clozapine and sulpiride but not haloperidol or olanzapine activate brain DNA demethylation. Proc Natl Acad Sci U S A. 2008;105(36):13614–9. https://doi.org/10.1073/pnas.0805493105.
43. Dong E, Chen Y, Gavin DP, Grayson DR, Guidotti A. Valproate induces DNA demethylation in nuclear extracts from adult mouse brain. Epigenetics. 2010;5(8):730–5. https://doi.org/10.4161/epi.5.8.13053.
44. Wang Q, Xu X, Li J, Liu J, Gu H, Zhang R, et al. Lithium, an anti-psychotic drug, greatly enhances the generation of induced pluripotent stem cells. Cell Res. 2011;21(10):1424–35. https://doi.org/10.1038/cr.2011.108.
45. Weaver IC, Cervoni N, Champagne FA, D'Alessio AC, Sharma S, Seckl JR, et al. Epigenetic programming by maternal behavior. Nat Neurosci. 2004;7(8):847–54.
46. Ma DK, Jang MH, Guo JU, Kitabatake Y, Chang ML, Pow-Anpongkul N, et al. Neuronal activity–induced Gadd45b promotes epigenetic DNA demethylation and adult neurogenesis. Science. 2009;323(5917):1074–7. https://doi.org/10.1126/science.1166859.
47. McGowan PO, Sasaki A, D'Alessio AC, Dymov S, Labonté B, Szyf M, et al. Epigenetic regulation of the glucocorticoid receptor in human brain associates with childhood abuse. Nat Neurosci. 2009;12(3):342–8. https://doi.org/10.1038/nn.2270.
48. Murgatroyd C, Patchev AV, Wu Y, Micale V, Bockmühl Y, Fischer D, et al. Dynamic DNA methylation programs persistent adverse effects of early-life stress. Nat Neurosci. 2009;12(12):1559–66. https://doi.org/10.1038/nn.2436.
49. Yehuda R, Bierer LM. The relevance of epigenetics to PTSD: implications for the DSM-V. J Trauma Stress. 2009;22(5):427–34. https://doi.org/10.1002/jts.20448.
50. Hunter RG, McEwen BS. Stress and anxiety across the lifespan: structural plasticity and epigenetic regulation. Epigenomics. 2013;5(2):177–94. https://doi.org/10.2217/epi.13.8.
51. Rusiecki JA, Chen L, Srikantan V, Zhang L, Yan L, Polin ML, Baccarelli A. DNA methylation in repetitive elements and post-traumatic stress disorder: a case-control study of US military service members. Epigenomics. 2012;4(1):29–40. https://doi.org/10.2217/epi.11.116.
52. Fraga MF, Ballestar E, Paz MF, Ropero S, Setien F, Ballestar ML, et al. Epigenetic differences arise during the lifetime of monozygotic twins. Proc Natl Acad Sci U S A. 2005;102(30):10604–9.
53. Miyake K, Yang C, Minakuchi Y, Ohori K, Soutome M, Hirasawa T, Kazuki Y, et al. Comparison of genomic and epigenomic expression in monozygotic twins discordant for Rett syndrome. PLoS One. 2013;8(6):e66729. https://doi.org/10.1371/journal.pone.0066729.
54. Silveira PP, Portella AK, Goldani MZ, Barbieri MA. Developmental origins of health and disease (DOHaD). J Pediatr. 2007;83(6):494–504.
55. Lillycrop KA, Phillips ES, Torrens C, Hanson MA, Jackson AA, Burdge GC. Feeding pregnant rats a protein-restricted diet persistently alters the methylation of specific cytosines in the hepatic PPAR alpha promoter of the offspring. Br J Nutr. 2008;100(2):278–82. https://doi.org/10.1017/S0007114507894438.
56. Tobi EW, Lumey LH, Talens RP, Kremer D, Putter H, Stein AD, et al. DNA methylation differences after exposure to prenatal famine are common and timing- and sex-specific. Hum Mol Genet. 2009;18(21):4046–53. https://doi.org/10.1093/hmg/ddp353.
57. Lim D, Bowdin SC, Tee L. Clinical and molecular genetic features of Beckwith-Wiedemann syndrome associated with assisted reproductive technologies. Hum Reprod. 2009;24(3):741–7. https://doi.org/10.1093/humrep/den406.
58. Bliek J, Alders M, Maas SM, Oostra RJ, Mackay DM, van der Lip K, et al. Lessons from BWS twins: complex maternal and paternal hypomethylation and a common source of haematopoietic stem cells. Eur J Hum Genet. 2009;17(12):1625–34. https://doi.org/10.1038/ejhg.2009.77.
59. Lambrot R, Xu C, Saint-Phar S, Chountalos G, Cohen T, Paquet M, et al. Low paternal dietary folate alters the mouse sperm epigenome and is associated with negative pregnancy outcomes. Nat Commun. 2013;4:2889. https://doi.org/10.1038/ncomms3889.
60. Popp C, Dean W, Feng S, Cokus SJ, Andrews S, Pellegrini M, et al. Genome-wide erasure of DNA methylation in mouse primordial germ cells is affected by AID deficiency. Nature. 2011;463(7284):1101–5. https://doi.org/10.1038/nature08829.
61. Daxinger L, Whitelaw E. Transgenerational epigenetic inheritance: more questions than answers. Genome Res. 2010;20(12):1623–8. https://doi.org/10.1101/gr.106138.110.
62. Horsthemke B. Heritable germline epimutations in humans. Nat Genet. 2007;39:573–4.
63. Chong S, Youngson NA, Whitelow E. Heritable germline epimutation is not the same as transgenerational epigenetic inheritance. Nat Genet. 2007;39(5):574–5.
64. Grossniklaus U, Kelly B, Ferguson-Smith AC, Pembrey M, Lindquist S. Transgenerational epigenetic inheritance: how important is it? Nat Rev Genet. 2013;14(3):228–35. https://doi.org/10.1038/nrg3435.
65. Law JA, Du J, Hale CJ, Feng S, Krajewski K, Palanca AM, et al. Polymerase IV occupancy at RNA-directed DNA methylation sites requires SHH1. Nature.

2013;498(7454):385–9. https://doi.org/10.1038/nature12178.

66. Khraiwesh B, Zhu JK, Zhu J. Role of miRNAs and siRNAs in biotic and abiotic stress responses of plants. Biochim Biophys Acta. 2012;1819(2):137–48. https://doi.org/10.1016/j.bbagrm.2011.05.001.
67. Becker C, Hagmann J, Müller J, Koenig D, Stegle O, Borgwardt K, et al. Spontaneous epigenetic variation in the Arabidopsis thaliana methylome. Nature. 2011;480(7376):245–9. https://doi.org/10.1038/nature10555.
68. Schmitz RJ, Schultz MD, Lewsey MG, O'Malley RC, Urich MA, Libiger O, Schork NJ, Ecker JR. Transgenerational epigenetic instability is a source of novel methylation variants. Science. 2011;334(6054):369–73. https://doi.org/10.1126/science.1212959.
69. Rakyan V, Whitelaw E. Transgenerational epigenetic inheritance. Curr Biol. 2003;13(1):R6.
70. Becker C, Weigel D. Epigenetic variation: origin and transgenerational inheritance. Curr Opin Plant Biol. 2012;15(5):562–7. https://doi.org/10.1016/j.pbi.2012.08.004.
71. Rakyan VK, Chong S, Champ ME, Cuthbert PC, Morgan HD, Luu KV, Whitelaw E. Transgenerational inheritance of epigenetic states at the murine Axin(Fu) allele occurs after maternal and paternal transmission. Proc Natl Acad Sci U S A. 2003;100(5):2538–43.
72. Relton CL, Davey Smith G. Two-step epigenetic Mendelian randomization: a strategy for establishing the causal role of epigenetic processes in pathways to disease. Int J Epidemiol. 2012;41(1):161–76. https://doi.org/10.1093/ije/dyr233.
73. Kappeler L, Meaney MJ. Epigenetics and parental effects. BioEssays. 2010;32(9):818–27. https://doi.org/10.1002/bies.201000015.
74. Waterland RA, Travisano M, Tahiliani KG. Diet-induced hypermethylation at agouti viable yellow is not inherited transgenerationally through the female. FASEB J. 2007;21(12):3380–5.
75. Anway MD, Cupp AS, Uzumcu M, Skinner MK. Epigenetic transgenerational actions of endocrine disruptors and male fertility. Science. 2005;308(5727):1466–9. Erratum in: Science. 2010;328(5979):690.
76. Manikkam M, Tracey R, Guerrero-Bosagna C, Skinner MK. Plastics derived endocrine disruptors (BPA, DEHP and DBP) induce epigenetic transgenerational inheritance of obesity, reproductive disease and sperm epimutations. PLoS One. 2013;8(1):e55387. https://doi.org/10.1371/journal.pone.0055387.
77. Seong KH, Li D, Shimizu H, Nakamura R, Ishii S. Inheritance of stress-induced, ATF-2-dependent epigenetic change. Cell. 2011;145(7):1049–61. https://doi.org/10.1016/j.cell.2011.05.029.
78. Franklin TB, Russig H, Weiss IC, Gräff J, Linder N, Michalon A, et al. Epigenetic transmission of the impact of early stress across generations. Biol Psychiatry. 2010;68(5):408–15. https://doi.org/10.1016/j.biopsych.2010.05.036.
79. Champagne FA, Weaver IC, Diorio J, Dymov S, Szyf M, Meaney MJ. Maternal care associated with methylation of the estrogen receptor-alpha1b promoter and estrogen receptor-alpha expression in the medial preoptic area of female offspring. Endocrinology. 2006;147(6):2909–15.
80. Crews D, Gillette R, Scarpino SV, Manikkam M, Savenkova MI, Skinner MK. Epigenetic transgenerational inheritance of altered stress responses. Proc Natl Acad Sci U S A. 2012;109(23):9143–8. https://doi.org/10.1073/pnas.1118514109.
81. Kondo M, Gray LJ, Pelka GJ, Christodoulou J, Tam PP, Hannan AJ. Environmental enrichment ameliorates a motor coordination deficit in a mouse model of Rett syndrome-Mecp2 gene dosage effects and BDNF expression. Eur J Neurosci. 2008;27(12):3342–50. https://doi.org/10.1111/j.1460-9568.2008.06305.x.
82. Lonetti G, Angelucci A, Morando L, Boggio EM, Giustetto M, Pizzorusso T. Early environmental enrichment moderates the behavioral and synaptic phenotype of MeCP2 null mice. Biol Psychiatry. 2010;67(7):657–65. https://doi.org/10.1016/j.biopsych.2009.12.022.
83. Nag N, Moriuchi JM, Peitzman CG, Ward BC, Kolodny NH, Berger-Sweeney JE. Environmental enrichment alters locomotor behaviour and ventricular volume in Mecp2 1lox mice. Behav Brain Res. 2009;196(1):44–8. https://doi.org/10.1016/j.bbr.2008.07.008.
84. Kerr B, Silva PA, Walz K, Young JI. Unconventional transcriptional response to environmental enrichment in a mouse model of Rett syndrome. PLoS One. 2010;5:e11534.
85. Guy J, Gan J, Selfridge J, Cobb S, Bird A. Reversal of neurological defects in a mouse model of Rett syndrome. Science. 2007;315(5815):1143–7.
86. Luikenhuis S, Giacometti E, Beard CF, Jaenisch R. Expression of MeCP2 in postmitotic neurons rescues Rett syndrome in mice. Proc Natl Acad Sci U S A. 2004;101(16):6033–8.
87. Lioy DT, Garg SK, Monaghan CE, Raber J, Foust KD, Kaspar BK, et al. A role for glia in the progression of Rett's syndrome. Nature. 2011;475(7357):497–500. https://doi.org/10.1038/nature10214.
88. Jablonka E, Raz G. Transgenerational epigenetic inheritance: prevalence, mechanisms, and implications for the study of heredity and evolution. Q Rev Biol. 2009;84(2):131–76.
89. Imura H. Life course health care and preemptive approach to non-communicable diseases. Proc Jpn Acad Ser B Phys Biol Sci. 2013;89(10):462–73.

MIX
Papier aus verantwortungsvollen Quellen
Paper from responsible sources
FSC® C105338

If you have any concerns about our products,
you can contact us on
ProductSafety@springernature.com

In case Publisher is established outside the EU,
the EU authorized representative is:
Springer Nature Customer Service Center GmbH
Europaplatz 3, 69115 Heidelberg, Germany

Printed by Libri Plureos GmbH
in Hamburg, Germany